Variational Principles of Continuum Mechanics with Engineering Applications

Volume 2: Introduction to Optimal Design Theory

Mathematics and Its Applications

Vadim Komkov

Air Force Institute of Technology,
Wright-Patterson Air Force Base, Ohio, U.S.A.

Variational Principles of Continuum Mechanics with Engineering Applications

Volume 2:
Introduction to Optimal Design Theory

D. Reidel Publishing Company

A MEMBER OF THE KLUWER ACADEMIC PUBLISHERS GROUP

Dordrecht / Boston / Lancaster / Tokyo

Library of Congress Cataloging in Publication Data

CIP

Komkov, Vadim.
Varational principles of continuum mechanics with engineering applications.

(Mathematics and its applications)
Includes bibliographies and indexes.
Contents: v. 1. Critical points theory—
v. 2. Introduction to optimal design theory.
1. Continuum mechanics. 2. Calculus of variations. I. Series: Mathematics and its applications (D. Reidel Publishing Company) II. Title.
QA808.2.K66 1986 531 85–28105
ISBN-13: 978-94-010-7791-0 e-ISBN-13: 978-94-009-2869-5
DOI: 10.1007/978-94-009-2869-5

Published by D. Reidel Publishing Company,
P.O. Box 17, 3300 AA Dordrecht, Holland.

Sold and distributed in the U.S.A. and Canada
by Kluwer Academic Publishers,
101 Philip Drive, Norwell, MA 02061, U.S.A.

In all other countries, sold and distributed
by Kluwer Academic Publishers Group,
P.O. Box 322, 3300 AH Dordrecht, Holland.

Softcover reprint of the hardcover 1st edition 1988

CONTENTS

SERIES EDITOR'S PREFACE

Approach your problems from the right end and begin with the answers. Then one day, perhaps you will find the final question.

'The Hermit Clad in Crane Feathers' in R. van Gulik's *The Chinese Maze Murders*.

It isn't that they can't see the solution. It is that they can't see the problem.

G.K. Chesterton. *The Scandal of Father Brown* 'The point of a Pin'.

Growing specialization and diversification have brought a host of monographs and textbooks on increasingly specialized topics. However, the "tree" of knowledge of mathematics and related fields does not grow only by putting forth new branches. It also happens, quite often in fact, that branches which were thought to be completely disparate are suddenly seen to be related.

Further, the kind and level of sophistication of mathematics applied in various sciences has changed drastically in recent years: measure theory is used (non-trivially) in regional and theoretical economics; algebraic geometry interacts with physics; the Minkowsky lemma, coding theory and the structure of water meet one another in packing and covering theory; quantum fields, crystal defects and mathematical programming profit from homotopy theory; Lie algebras are relevant to filtering; and prediction and electrical engineering can use Stein spaces. And in addition to this there are such new emerging subdisciplines as "experimental mathematics", "CFD", "completely integrable systems", "chaos, synergetics and large-scale order", which are almost impossible to fit into the existing classification schemes. They draw upon widely different sections of mathematics. This programme, Mathematics and Its Applications, is devoted to new emerging (sub)disciplines and to such (new) interrelations as exempla gratia:

- a central concept which plays an important role in several different mathematical and/or scientific specialized areas;
- new applications of the results and ideas from one area of scientific endeavour into another;
- influences which the results, problems and concepts of one field of enquiry have and have had on the development of another.

The Mathematics and Its Applications programme tries to make available a careful selection of books which fit the philosophy outlined above. With such books, which are stimulating rather than definitive, intriguing rather than encyclopaedic, we hope to contribute something towards better communication among the practitioners in diversified fields.

In the series editor's preface to the first volume of this two volume work I wrote that the time-gap between theoretical developments and applications of these was fast disappearing in many fields, including the part of mechanical engineering involving continuum mechanics. (Though that is definitely not the only part of mechanical engineering where this is the case; stochastics mechanics, for example, is another (cf. the book by P. Krée and Chr. Soize, Mathematics of random phenomena, in this series) and so is the part centering around questions related to identification and filtering).

Variational problems and optimization questions in continuum mechanics tend to involve a functional, a domain, exterior forces (or controls) and a PDE for the function for which an optimum (extremum) is sought. The more classical problems ask for the optimizing function. Other more modern questions ask for optimal exterior forces in some sense, or optimal shape of the domain involved, or such questions as how much of the boundary must be available (for control through boundary conditions) to be able, say, to control the vibrations of a satellite.

This book is mainly concerned with optimal shape and optimal exterior forces type problems.

All such problems tend to involve abstract differentiation in all kinds of infinite-dimensional (function) spaces and it is definitely not true that 'straightforward' Fréchet or Gateaux differentiation will lead to the right kind of numerical algorithms. A lot of modern mathematics is needed including substantial amounts from that again flowering area of research: the effective use of symmetry properties. (For the matter, also such things as nonstandard analysis have their applications to problems of optimal shape as the author has shown.)

Thus the mechanical engineer is faced with the problem that there are many sophisticated mathematical tools ready to be applied and the mathematician is confronted with the fact there are many important unsolved problems coming out of continuum mechanics. I expect this book will be most useful for both.

The unreasonable effectiveness of mathematics in science ...

Eugene Wigner

Well, if you know of a better 'ole, go to it.

Bruce Bairnsfather

What is now proved was once only imagined.

William Blake

As long as algebra and geometry proceeded along separate paths, their advance was slow and their applications limited.

But when these sciences joined company they drew from each other fresh vitality and thenceforward marched on at a rapid pace towards perfection.

Joseph Louis Lagrange.

Bussum, November 1987

Michiel Hazewinkel

Introduction to Volume 2

VARIATIONAL PRINCIPLES OF CONTINUUM MECHANICS, INTRODUCTION TO OPTIMAL DESIGN THEORY

The study of variational problems usually starts with the optimization of some functional $J(f)$, where f belongs to a class of admissible functions defined in a fixed region Ω of a Euclidean space and obeys certain a-priori specified constraints. The problem of minimizing $J(f(\underset{\sim}{x}))$, $x \in \Omega$, is equivalent to finding a solu-ution $\hat{f}(x)$ to an equation $L(f) = q$ where L is, generally a differential operator and q is a known function. More frequently problems modeled by some differential equations are formulated in a "variational form". Instead of "solving" a differential system one can attempt to find a function that assigns an extremal value to a functional.

Problems of this type restated in the form: minimize a functional $J(f)$: $H \to \mathbb{R}$, where f is allowed to vary in a class of admissible functions H, belong to calculus of variations.

The restatement of laws of continuum mechanics in a variational form, which was the principal topic of Volume 1 of this work, has been the subject of intensive investigation for almost two centuries, with some of most illustrious names in mathematics and physics associated with it. However, the classical formulation of problems in the calculus of variations is only one of the possible problems arising in the optimization of functionals. Let us consider a specific example. The Saint Venant's problem of pure torsion is modeled by the equation

$$\Delta \Phi = f(x,y),\ x,\ y \in \Omega \subset \mathbb{R}^2, \Delta \equiv \partial^2/\partial x^2 + \partial^2/\partial y^2,$$

where Ω is the two-dimensional compact region

occupied by the shaft's cross-sectional area. $\Phi(x,y)$ is the stress function, such that

$$\frac{\partial \Phi}{\partial x} = \tau_{yz}, \quad \frac{\partial \Phi}{\partial y} = -\tau_{xz}$$

(τ_{yz}, τ_{xz} are the shear stress components, in the usual engineering notation), with all other stress components assumed to be identically equal to zero. Φ vanishes on the boundary, i.e. $\Phi|_\Gamma \equiv 0$.

In the usual formulation of the Saint Venant's problem $f(x,y) \equiv$ constant, the constant is chosen to be (-2) for convenience. The same problem arises in the modeling of the static deflection of a membrane. If Ω denotes the region occupied by the membrane in the x,y - plane, $f(x,y)$ - the pressure, $\rho(x,y)$ - the mass per unit area, the potential energy assumes the form

$$V = \int_\Omega \rho(x,y) \cdot |\text{grad } \Phi(x,y)|^2 dxdy - \int_\Omega (f\Phi)dxdy, \tag{0.2}$$

where $\Phi(x,y)$ denotes the deflection of the membrane in the z-direction (i.e. perpendicular to the x,y - plane).

The corresponding Euler-Lagrange system is

$$\begin{cases} -\text{grad } (\rho \cdot \text{grad } \Phi) = f, \\ \Phi|_\Gamma \equiv 0. \end{cases}$$

We see that a number of problems can be formulated.

a) The classical problem of the calculus of variations: Find a function Φ in the Sobolëv space $H^1_0(\Omega)$ that minimizes the functional (0.2)

for a given distribution of pressure $f(x,y)$ and a given shape of the simply connected domain Ω (therefore for a given boundary Γ).

b) Subject to some constraints, such as

$\int_{\Omega} f(x,y)\,dxdy = 1$, and $f_1 < |f| < f_2$, find

$f(x,y)$ in some admissible class F, such that $f(x,y)$ assigns an optimal value to the functional $V(\Phi)$. Here Ω and $\rho(x,y)$ are given.

c) Given $f(x,y)$, and the domain Ω find $\rho(x,y)$ such that $V(\Phi(\rho))$ assumes an optimal value.

(d) Given $\rho(x,y)$, $f(x,y)$ in some region C, with measure(C) ≥ 1, find $\Omega \subset C$, with

$\int_{C-\Omega} dxdy = 1$, such that $V(\Phi,\Omega)$ assumes

optimal value.

We have been deliberately vague about the optimality requirements. "Optimal" could mean minimal, maximal, "close" to some specific value. $V(\Phi,\Omega)$ does not have to be the energy, and its optimum could mean that Φ approximates in some sense a given deflection, or that the deflection at some specific points assumes preassigned values, or a number of other "optimality" criteria.

The problem (a) was designated as a classical problem. It is the problem discussed extensively in almost any Calculus of Variations text.

Problem (b) involves optimization of the forcing (or control) term. What external forces do we need to apply to the membrane to optimize the cost functional?

If we replace the static problem by a dynamic one (vibration of a membrane) we can find an abundance of modern literature (i.e. after 1945) dealing with this class of problems. It is one of the basic problems of modern control theory.

Problem (c) deals with design of the thickness or mass distribution of the membrane. Some authors refer to such problems as "control of the coefficients." Rapid progress in understanding the problem (c) was made in the late 70-s and the 80-s. The continuous dependence of energy on coefficients regarded as vectors in a Hilbert space turned out to be true in most cases and false in some. Thus, a straightforward Fréchet or, Gateaux differentiation sometimes could produce a convergent numerical algorithm, and at other times could lead to obviously incorrect designs. Problem (d) is probably the least investigated and beast amenable to purely heuristic manipulations. Recently it became a "red hot" item of research, with the "French school" taking the lead in advancing the 1910 idea of Hadamard. It involved variation of the Green's function for the problem, due to a small perturbation of the domain in a manner similar to the vanishing of the first variation in problem (a).

The problems of the type (c) and (d) are the primary subject of this volume. All problems (a)-(d) are usually attacked by techniques that can be regarded as abstract differentiation . The value of the functional is extreme if either the variable quantity lies on the boundary of the admissible region in appropriately chosen space, or else if some abstract first derivative vanishes. These two possibilities are exactly mirrored in some maximal principles (Pontryagin's theory), bang-bang principles in the former case or else in the Gauss-Hertz principle, or the zero sensitivity postulate, or in Hadamard's formulation for the Green's function in the latter case. Thus, we are able to unify several seemingly disconnected ideas and to reexamine critically the corresponding numerical schemes. Moreover, some common features of all of the problems labelled (a) - (d) become quite clear, and the difficulties also appear to have some common origins and are generally related to the lack of smoothness or to the poor choice of what is "admissible".

Chapter 1

Changes of Coordinates and Variation of the Coefficients

1.1 The state space.

Problems of engineering design involve a "cost functional", constraints, and admissibility considerations.

In principle, at least, we wish to design some mechanical or structural project as cheaply as possible, minimizing "a cost functional". We must obey some rules, that is constraints, some laws of physics, which we have no power to alter, and some manufacturing limitations. Also we need to comply with some specifications. For example the bridge must be able to withstand reasonable traffic and wind loads, a machine must be able to operate for a reasonable period of time without excessive wear, a circuit must be able to withstand an unexpected surge of current. Thus, we have a number of constraints imposed on the optimization problem.

Finally, we must decide the admissibility of the mathematical model. But a first step in any modeling must be the decision regarding the choice of coordinates and the mathematical description of the "state of the system".

The underlying frame of reference is commonly based on Euclidean space $\mathbf{R}^3$ with Cartesian coordinates $\underset{\sim}{x} = \{x, y, z\}$ or $\underset{\sim}{x} = \{x_1, x_2, x_3\}$ with the undefined concept of points in that space. A mechanical system (or a continuum) is said to have n-degrees of freedom if its configuration or state is completely defined by n- independent coordinates $\{q^i\}$, $i = 1, 2, \ldots n$. We assume that "admissible" variables of system $\{q^i\}$ form a local coordinate cover. At each point $\tilde{q}$ in the configuration space there is a neighborhood of $\tilde{q}$ spanned by $\{q^i\}$, which is locally Euclidean. This means that there is a mapping from the neighborhood $N_{\tilde{q}}$ of $\tilde{q}$ into some neighborhood of zero in the Euclidean space that is an isomorphism (it is one to one with a unique inverse). The configuration space with n-degrees of freedom, or the state space is a set $S \subset \mathbb{R}^n$ with a local coordinate cover $\{q^i\}$, $i = 1,2,...n$; (In fact S is a manifold.)

The Kinematic event space is a subset of $\mathbb{R} \times S$, that is an ordered pair $\{t, q^i(t)\}$, with t interpreted as time and $\{q^i(t)\}$ as the state of the system at time t.

We assume Newton's rather than Einstein's or even more recent interpretation of the concept of time. All events are well ordered with respect to time axis, which is an isomorph of the real line. Two separate events are always universally ordered with respect to all observers. Either event one preceeds event two, or they are simultaneous, or else event two preceeds event one. Time and state space are independent. Motion of a particle (that is of a single point) is described by a parameterized path : $\underset{\sim}{x} = \underset{\sim}{x}(t)$, $t \geq t_o$, or a map $t(\in \mathbb{R}) \to \mathbb{R}^n$. We insist that this map is defined for either a discrete system (with n-degrees of freedom) or for a continuum.

1.2 A change of coordinates.

The study of motion of "points" in a continuum, may consist in effect of taking a ride on point particles of the continuum in its motion.

Let $\underset{\sim}{\xi}(t_o)$ be the position of some arbitrary point of the continuum at some (call it initial) time t_o. We regard the collection of particle paths as a system evolving from its initial state at t_o. The position at time t is given by the relation

(1.1) $\underset{\sim}{x} = \underset{\sim}{x}(\underset{\sim}{\xi}, t)$ with $\underset{\sim}{x}(\underset{\sim}{\xi}, t_o) = \underset{\sim}{\xi}$.

$\underset{\sim}{x}$ is regarded as a function of $\underset{\sim}{\xi}$ and t, while

the time t is indentified with an infinite ray of the real line.

The state space S is a subset of $\mathbb{R}^n$, or of a Hilbert space $\mathbb{H}$. Without any loss of generality we can assume $t_o = 0$, that is identify the time ray with $\mathbb{R}_+$.

We insist that the map $t \to S \subset \mathbb{R}^n$ or $t \to S \subset \mathbb{H}$ is defined.

For each instant $t \in \mathbb{R}_+$ we have only one possible configuration $\{\underset{\sim}{q}\} \in S$ of the system. We refer to this uniqueness property by calling the system deterministic. Knowing $q(0)$ we can determine (at least in principle) the unique state of the system at any future time $t > t_o$. The motion of each "point" or "particle" $\underset{\sim}{x}(t) \in \mathbb{R}^3$ follows that point according to an equation

(1.1) $\underset{\sim}{x} = \underset{\sim}{x}(\underset{\sim}{\xi}, t)$, $\underset{\sim}{x}(\underset{\sim}{\xi}, 0) = \underset{\sim}{\xi}$

If each particle's path can be "retracted", that is if we can uniquely solve for $\underset{\sim}{\xi}$, given the relation $\underset{\sim}{x}(\underset{\sim}{\xi}, t) = \hat{\underset{\sim}{x}}(t)$ about the position $\hat{\underset{\sim}{x}}$ at time t, we call this property the <u>solenoidal</u> property, or the <u>impenetrability</u>. We assert that two paths cannot cross each other, and only one particle can occupy some point $\underset{\sim}{x}$ in the Euclidean space at a given time.

We can regard $\underset{\sim}{x}$, t as space variables $\underset{\sim}{\xi}$, t as material variables. It is unfortunate that we perpetuate a confused notation in which $\underset{\sim}{x}$ sometimes denotes a coordinate system and sometimes a function of time, when we identify $\underset{\sim}{x}$

with the position of a particle in the coordinate system designated by the same symbol, and sometimes an ordered n-tuple corresponding to a specific evaluation of the function $t \to \underset{\sim}{x}(t)$.

We discuss possible ways of clarifying this confusing notation in an appendix. For the time-being we shall perpetuate the mess created by our predecessors, where the usual notation fails to distinguish between a function and its range.

For a point-particle we specify the mass density by introducing the Dirac delta measure. (Read [1], [2], or [3]).

$$(1.2)\begin{cases} \rho(\underset{\sim}{x}, t) = \rho_0 \cdot \delta(\underset{\sim}{x}(t) - \underset{\sim}{x}_0) \\ \rho(\underset{\sim}{x}(\underset{\sim}{\xi}), t) = \rho_0 \cdot \delta(\underset{\sim}{x}(\underset{\sim}{\xi}) - \underset{\sim}{x}_0(\xi)) \end{cases}$$

Here $\{\underset{\sim}{x}, t\}$ - are spatial independent variables

$\{\underset{\sim}{\xi}, t\}$ - are "material" coordinates, or material independent variables.

ρ_0 is some constant associated with the mass of the particle in a suitably chosen system of (physical) units. The "point in space" and "position of a particle" are interchangable concepts, corresponding to the coordinate transformation (1.1):

$$\underset{\sim}{x} = \underset{\sim}{x}(\underset{\sim}{\xi}, t).$$

The density of a mass at a point, or the mass density function is given

$$\rho(\underset{\sim}{x}(\underset{\sim}{\xi}, t)) = \mu(\underset{\sim}{\xi}, t)$$

that could be a "genuine" (that is locally integrable) function, or it could be a generalized function (for example Dirac delta, or its derivative). Two concepts are of primary importance in continuum mechanics: the position and all of its pertinent derivatives, and the mass density and its derivatives.

We can regard the position of a particle at

$\underset{\sim}{x}$ and its velocity $\mathbf{V}(\underset{\sim}{x},t)$ as basic independent quantities. The validity of spatial-material coordinate transformation implies

(1.4) $\underset{\sim}{v}(\underset{\sim}{x},t) = \underset{\sim}{v}(\underset{\sim}{x}(\xi, t), t) = \mathbf{V}(\underset{\sim}{\xi}, t).$

The chain rule implies that

(1.5) $$\frac{\partial \mu(\underset{\sim}{\xi}, t)}{\partial t} = \sum_{i=1}^{3} \frac{\partial \rho}{\partial x^i} \cdot \frac{\partial x^i}{\partial t}\bigg|_{x^i = x^i(\underset{\sim}{\xi},t)}$$

$$= \sum \frac{\partial \rho(\underset{\sim}{x}, t)}{\partial x^i} . V^i(\underset{\sim}{\xi}, t)$$

On the other hand

(1.6) $$\frac{\partial \mu(\xi, t)}{\partial t}\bigg|_{\underset{\sim}{\xi} = \underset{\sim}{\xi}(\underset{\sim}{x}, t)} = \sum \frac{\partial \rho(\underset{\sim}{x}, t)}{\partial x^i} . V^i(\underset{\sim}{x},t)$$

$$+ \frac{\partial \rho(\underset{\sim}{x}, t)}{\partial t} = \frac{D\rho(\underset{\sim}{x}, t)}{Dt}, \text{ or}$$

(1.7) $$\frac{\partial \mu(\xi, t)}{\partial t}\bigg|_{\underset{\sim}{\xi} = \underset{\sim}{\xi}(\underset{\sim}{x}, t)} \equiv \frac{D\rho(\underset{\sim}{x}, t)}{Dt}$$

$$= \frac{\partial \rho(\underset{\sim}{x},t)}{\partial t} + (\underset{\sim}{v}(\underset{\sim}{x},t) \cdot \nabla \rho(\underset{\sim}{x},t)),$$

where $\cdot$ denotes the usual dot product in three dimensions.

1.3 Conservation of mass, conservation of energy.

If in a deformation process we observe the evolution of some bounded region Ω_0 occupied by the material, that is bounded at time $t_0 = 0$, so that tracing the motion of each point of we associate bounded open regions Ω_t with each time instant t. The mass conservation property is expressed by the equation

$$(1.8)\quad \frac{d}{dt}\left\{ \iiint_{\Omega_t} \rho(\underset{\sim}{x}, t)\cdot d\underset{\sim}{x} \right\} = 0.$$

In fact any "conservation" property assumes an identical form. If, for example, $e(\underset{\sim}{x}, t)$ is some form of energy density, then the energy conservation law is given by

$$(1.9)\quad \frac{d}{dt}\left\{ \iiint_{\Omega_t} e(\underset{\sim}{x}, t)\, d\underset{\sim}{x} \right\} = 0.$$

The mapping $t \overset{\mathcal{D}}{\to} \Omega_t$ $(\mathbb{R} \to \mathbb{R}^3)$, $\Omega(t) = \Omega_t$, $\Omega(0) = \Omega_o$ is assumed to be continuous with respect to the norm $\|\cdot\|$, where $\|\Omega_1 - \Omega_2\| = \sup. \|\underset{\sim}{x} - \underset{\sim}{y}\|$, $\underset{\sim}{x} \in \Omega_1$, $\underset{\sim}{y} \in \Omega_2$. It is easy to check that it is a norm. Since $\|\Omega\| = \sup_{\underset{\sim}{x}\in\Omega} \|\underset{\sim}{x}\|$, obviously $\|c\,\Omega\| = |c|\;\|\Omega\|$ for any $c \in \mathbb{R}$, and $\|\Omega\| > 0$ if and only if $\Omega \neq \emptyset$, and the triangular inequality is satisfied. Therefore, continuity of the map $t \to \Omega_t$ is defined. (Given $\varepsilon > 0$ there exists $\delta > 0$ such that $\|\Omega_{t_1} - \Omega_{t_2}\| < \varepsilon$ whenever $|t_1 - t_2| < \delta$.)

We refer to a functional $I(\Omega_t)$ as the invariant of motion if $\frac{dI(\Omega_t)}{dt} \equiv 0$. For example, the total mass contained in the region Ω_t is invariant if

$$(1.10)\quad \frac{d}{dt}\iiint_{\Omega_t} \rho(\underset{\sim}{x},t)\, d\underset{\sim}{x} = \frac{d}{dt}\, I(\Omega_t) \equiv 0,$$

for all $t \in \mathbb{R}_+$.

Refering the deformation process to the material coordinates we derive

$$\text{(1.11)} \quad \frac{d\, I\,(\Omega_t)}{dt} = \iiint_{\Omega_o} \Big[\frac{\partial \mu(\xi,t)}{\partial t}\, J(x/\xi,(t)) + \mu(\xi,t)\,\frac{\partial J(\underset{\sim}{x}/\underset{\sim}{\xi},\ (t))}{\partial t}\Big]\, d\underset{\sim}{\xi}\ ,\quad \text{where}$$

$J(\underset{\sim}{x}/\underset{\sim}{\xi},\ (t))$ is the time-dependent Jacobian

$$J(x/\xi\,,(t)) = \Big|\frac{\partial x_i(\underset{\sim}{\xi})}{\partial \xi_j}\Big| .$$

An intuitive meaning of the Jacobian $J(x/\xi)$ is as follows: $J(x/\xi)$ is a function of time representing the ratio of volume of an infinitesimal element that occupies the volume $d\underset{\sim}{x}$ compared to the initial volume (at time $t_o = 0$) at position ξ. The Euler's formula (sometimes called Euler's expansion formula) relates the rate of change of the Jacobian $J(\underset{\sim}{x}/\underset{\sim}{\xi}\,,(t))$ to the divergence of the velocity vector.

$$\frac{\partial J(x/\xi)}{\partial t} = \big[\nabla_{\underset{\sim}{x}} \cdot \underset{\sim}{v}\ (\underset{\sim}{x})\big]\cdot J\ (x/\xi),$$

where $\nabla_x \cdot \underset{\sim}{v}(\underset{\sim}{x}) = \frac{\partial v_1(\underset{\sim}{x})}{\partial x_1} + \frac{\partial v_2(\underset{\sim}{x})}{\partial x_2} + \frac{\partial v_3(\underset{\sim}{x})}{\partial x_3}$.

Two so called "fundamental theorems of calculus of variations" were stated by du Bois-Raymond. Let us restate them in a slightly more modern terminology.

<u>Theorem 1.1</u> If in a given domain Ω, the functional relation: $\langle f,g\rangle_\Omega = \iiint_\Omega [f(\underset{\sim}{x})\cdot g(\underset{\sim}{x})]\, d\underset{\sim}{x} = 0$ is true for any $g(x) \in L_2(\Omega)$, (or only for any

$g(\underset{\sim}{x})$ in a set of functions G that is complete in Ω), then $f(\underset{\sim}{x}) = 0$ almost everywhere in Ω.

Note: a set $G \subset L_2(\Omega)$ is complete in $L_2(\Omega)$ if for any $g \in G$, $\langle h,g \rangle = 0$ implies that $h \in G$. That is, any function orthogonal to g is in G.

Theorem 1.2. If $\iiint_{\Omega_i} f(\underset{\sim}{x})\, d\underset{\sim}{x} = 0$, $f \in L_2(\Omega)$ for any Lebesgue measurable subset $\Omega_i \subset \Omega \subset R^3$, then $f(\underset{\sim}{x}) \equiv 0$ almost everywhere in Ω. Theorems 1.1 and 1.2 can be simplified if we can assume that $f(\underset{\sim}{x})$ is a continuous function of $\underset{\sim}{x}$ in Ω. Specifically, we can replace "almost everywhere in Ω" by "everywhere in Ω".

A simple consequence of theorem 1.2 combined with Euler's expansion formula is the fundamental mass conservation equation of Bernoulli and Euler.

$$(1.12) \quad \frac{D\rho(\underset{\sim}{x})}{Dt} + \rho(\nabla_x \cdot \underset{\sim}{v}) = \frac{\partial \rho(\underset{\sim}{x}(t))}{\partial t} + \nabla_x \cdot (\rho \underset{\sim}{v}) \equiv 0.$$

That is a restatement of equations (1.10) and (1.11).

One can also establish a transformation between the acceleration components expressed in spatial and material coordinates

$$\left.\frac{\partial \underset{\sim}{V}(\underset{\sim}{\xi},t)}{\partial t}\right|_{\underset{\sim}{\xi} = \underset{\sim}{\xi}(\underset{\sim}{x},t)} = \frac{D\underset{\sim}{v}(\underset{\sim}{x},t)}{Dt} = \frac{\partial \underset{\sim}{v}}{\partial t} + (\underset{\sim}{v} \cdot \nabla)\underset{\sim}{v}.$$

Let us suppose that the

region $\Omega_o \subset \mathbb{R}^3$ occupied by a material, for example, an elastic body, is deformed.

$$t : \Omega(t = 0) = \Omega_o \to \Omega(t),$$

so that a region $\Omega(t)$ represents the shape of the body (or material) at time t, while Ω_o is the initial shape. In the incompressible fluid flow problems, we can identify certain region Ω_o at $t = 0$ within the fluid and follow the trajectory of each point within the fluid identifying regions $\Omega(t)$. The initial coordinates ξ are related to spatial coordinates $\underset{\sim}{x}(\underset{\sim}{\xi}, t)$ by a one to one invertible transformation.

The laws of differentiation under the integral sign are given by various versions of the abstract Stoke's theorem. The simplest form is a generalization of the mass conservation principle. We replace

$$\frac{d}{dt} \iiint_{\Omega(t)} \rho(\underset{\sim}{x}, t)\, d\underset{\sim}{x} = 0 \qquad \text{by}$$

$$(1.10^a) \quad \frac{d}{dt} \iiint_{\Omega(t)} \rho(\underset{\sim}{x}, t)\, d\underset{\sim}{x} = -\iint_{\partial\Omega(t)} (\rho \underset{\sim}{v}(\underset{\sim}{x},t) \cdot \mathbf{n})\, d\underset{\sim}{s},$$

where $\partial\Omega$ denotes the boundary of Ω, $\mathbf{n}$ is the unit vector normal to $\partial\Omega$, $v(\underset{\sim}{x},t)$ is the velocity expressed in terms of the spatial coordinates $\{\underset{\sim}{x}\}$, and $d\underset{\sim}{s}$ is the measure of surface area assigned to the surface $\partial\Omega$. The equation (1.10^a) compares the rate of change of the mass contained in $\Omega(t)$ to the net mass flux across the boundary $\partial\Omega(t)$.

Here we assume that everything is properly defined. The derivatives exist, the boundary has a unique normal at every point and no ugly questions have to be answered.

Applying Gauss' divergence theorem to the boundary term in (1.10^a) we recover the equation

$$(1.10^b)\quad \iiint_{\Omega(t)} \left[\frac{\partial \rho}{\partial t} + \nabla \cdot (\rho \underset{\sim}{v}) \right] d\underset{\sim}{x} = 0 \ .$$

Since this is true for any region $\Omega(t)$, the second lemma of du Bois Raymond implies that the integrand is identically equal to ,(or almost everywhere in Ω is equal to)zero,if we stop short of assuming continuity of the derivatives inside the integral sign in (1.10^b). Thus,we rederive the basic equation (1.12)

1.3^a Differential form of conservation laws

Suppose that ρ, X are functions that satisfy the relation

$$(1.13)\quad \frac{\partial \rho}{\partial t} + \frac{\partial X}{\partial x} = 0 \quad \text{in Cartesian coordinates :}$$

(t,x).

This is the case when $X = \rho(x,t) \cdot v(x,t)$. Then the equation (1.13) is integrated as the Bernoulli-Euler equation of continuity (that is of conservation of mass):

$$(1.13^a)\quad \frac{\partial \rho}{\partial t} + \frac{\partial(\rho v)}{\partial x} = 0 \, . \quad \text{(See [4], or [8].)}$$

A similar equation describes the conservation of the total electric charge, or of some form of energy.

The Korteweg - de Vries (K. deV.) equation describing two-dimensional weakly nonlinear long waves in an inviscid and incompressible fluid is of the same form

$$(1.14)\quad \frac{\partial u}{\partial t} + \frac{\partial}{\partial x} \left(-3u^2 + u_{xx}\right) = 0 \ .$$

The conserved quantity in the equation (1.13) is

$$\int_{\mathbb{R}} \left[\frac{\partial \rho}{\partial t}\right] dx = [- X \,]_{+\infty}^{+\infty} = 0, \qquad \text{or}$$

$$\frac{d}{dt} \int_{-\infty}^{+\infty} [\, \rho(x,t) \,] \, dx = 0, \qquad \text{and}$$

(1.15) $\int_{-\infty}^{+\infty} \rho \, dx = \text{constant.}$

Thus, in the Korteweg - de Vries equation the conserved quantity is

$\int_{-\infty}^{+\infty} [u(x,t)]dx$, where $u(x,t)$ represents the depth below the free surface of the liquid.

Other conservation laws are easily derived by observing the K.-deV. equation may be rewritten as

$$\frac{\partial}{\partial t} (\tfrac{1}{2} u^2) + \frac{\partial}{\partial x} (-2 u^3 + u \cdot u_{xx} - \tfrac{1}{2} u_x{}^2) = 0 .$$

See [4].

Thus, $\int_{-\infty}^{+\infty} (u^2) \, dx = \text{constant}$

is another conservation law.

Miura, Gardner and Kruskal, constructed a scheme for producing an infinite sequence of such conservation laws. (See [5] and [6].

Another example of equation of the form (1.13) is the Burger's equation

(1.16) $u_t + u \cdot u_x - \nu u_{xx} = 0$,

which can be rewritten as

(1.16^a) $\frac{\partial u}{\partial t} + \frac{\partial}{\partial x} (\tfrac{1}{2} u^2 - \nu u_x) = 0.$

Therefore $u(x,t)$ plays the same role in (1.16) as $\rho(x,t)$ in the conservation principle (1.15).

Further examples:

A solitary wave in an elastic medium is modeled by the equation

(1.17) $u_{tt} = u_{xx} + u_x u_{xx} + u_{xxxx}$,

or

(1.17^a) $\frac{\partial u_t}{\partial t} = \frac{\partial}{\partial x} [\tfrac{1}{2} (u_x)^2 + u_x + u_{xxx}]$

(See Toda [7]). It is easy to recognize the conserved quantity

Chapter 1

References for Chapter 1

[1] I.M. Gel'fand and G.E. Shilov, Volume I of Generalized Functions, Academic Press, New York, 1964. Translation from Russian by George Saletan.

[2] A. Zemanian, Distribution theory and transform analysis, McGraw Hill, New York, 1965.

[3] L. Schwartz, Theory des distributions, Vol. I and II, Herman & Cie, Paris, 1957 and 1959.

[4] P.G. Drazin, Solitons, Cambridge University Press, London Mathematical Society. Lecture Series #85, Cambridge University Press, 1983.

[5] R. M. Miura, C.S. Gardner and M.D. Kruskal, Kerteweg-de Vries equation ang generalizations, J. Math. Physics, #9, 1968, p. 1204-1209.

[6] R.M. Miura, The Korteweg-de Vries equation. A survey of results, S.I.A.M. Review, 18, (1976), p. 412-459.

[7] M. Toda, Wave propargation in anharmonic lattice, J. Phys. Soc. Japan, 23, (1967), p. 501-506.

[8] N.H. Ibrahimov, Invariant variational problems and conservation laws, Teoret. i. Mat. Fizika, Vol. 1, #3, 1969, p. 350-369.

Chapter 2

Group Theoretic Methods with Applications to Continuum Mechanics and Noether's Theory

2.1 Introduction to groups of transformation.

Let $\mathcal{M}$ be a manifold. All transformations $\mathcal{M} \to \mathcal{M}$ (i.e. 1:1 invertible mappings) form a group with composition of transformations as the group operation o. Let $\theta(\mathcal{M})$ denote the group of transformations on $\mathcal{M}$.

Let G be a group. We should really write it as (G,·) where · is the group operation). Π: a group isomorphism between G and the subgroup of the group of transformations $\theta(\mathcal{M})$. Then Π is called the representation of the group **G** in $\theta(M)$, i.e. $\Pi: (\mathbf{G},\cdot) \to (\theta(\mathcal{M}),\bullet)$, $\Pi(a\bullet b) = \Pi(a) \bullet \Pi(b)$, where $a,b \in \mathbf{G}$, $\cdot$ is the group operation in **G** and $\bullet$ is the composition of transformations of $\mathcal{M}$.

Parameterized groups. In this discussion we shall restrict the parameters to real numbers only. A k-parameter group of transformations $\theta(\mathcal{M})_k$ is a representation of the Cartesian product $\mathbb{R}^k$ of k-groups of real numbers under addition ($\mathbb{R}$, +) into $\theta(\mathcal{M})$. The group representation is the map $\Pi: \mathbb{R}^k \to \theta(\mathcal{M})$.

As an example, consider addition of real numbers and translations along the real axis. In this case the manifold $\mathcal{M}$ is the real line, $\theta_a(\mathcal{M})$ is a translation of $\mathcal{M}$ assigning to each point $x \in \mathcal{M}$ the point $x_a = x + a$. The isomorphism between the group of real numbers under addition and the translation of $\mathcal{M}$ is trivial.

Similarly, the group of translations of $\mathcal{M} = \mathbb{R}^2$ is isomorphic to the group of complex numbers under addition. Regarding the complex plane as a direct product of $\mathbb{R} \times \mathbb{R}$(ordered pairs of reals), with the usual addition, we have again a trivial isomorphism between additions and $\mathbb{C}$ and translations in M.

We also require that $\Pi(0,0,\ldots 0) = I_{\mathcal{M}}$. That is, the origin in $\mathbb{R}^k$ is mapped into the identity transformation of $\mathcal{M}$. Generally $\mathcal{M}$ is either a Hilbert or only a Banach space, or an appropriate subset of such a space.

A one parameter group of transformations is a map $\Pi: \mathbb{R} \to \theta(\mathcal{M})$ $\Pi(a) = \theta_a(\mathcal{M})$, $a \in \mathbb{R}$.

If we wish to evaluate the transformation θ at the point $x \in \mathcal{M}$, we compute $\theta_a(x) = \Pi(a,\underset{\sim}{x})$ for $x \in \mathcal{M}$. We could regard the action of the group of real numbers under addition: $G_R = (\mathbb{R},+)$ on $\mathcal{M}$ as a pointwise operation

$$\Pi(a,x) = \Pi_a(\underset{\sim}{x}) = \theta_a(\underset{\sim}{x}).$$

Because G_R is a group of real numbers under addition, we inherit all properties of this group in the properties of the group $\theta_a(\mathcal{M})$ of representations of G_R in $\theta(M)$.

Groups discussed here are Lie groups. For an introduction to the theory of Lie groups and Lie algebras read for example Jacobson [10] Vol. III. For an advanced presentation see Bourbaki [11].

In general, a "local" theory represents a variational formulation. Some nonlinear phenomena cannot be defined on the product space $\mathcal{M} \times \mathbb{R}$ where $\mathcal{M}$ is the kinematic manifold and $\mathbb{R}$ the real line representing all possible values of a parameter. Generally the values of the parameter make physical sense only on some subset Δ of $\mathbb{R}$. A local Lie group $\mathbf{G}$ of transformations on $\mathcal{M}$ must obey the following axioms; $\phi \in G$ implies that it is a map $\Phi: V \times \Delta \to \mathcal{M}$, where $V \subset \mathcal{M}$ is such that $x \in M$, and $\forall\, a \in \Delta \Rightarrow \phi(x,a)$ is defined and there exists an (open) neighborhood of x, $N_x \subset V \subset \mathcal{M}$, for which $\phi: N_x \times \Delta$ is defined.

Also $\phi(x,0) = x \qquad \forall x \in V$,

and $\phi \in C^\infty(V \times \Delta)$.

$$\phi(\phi(x,a),b) = \phi(x, a+b) \text{ if } a, b \in \mathbb{R} \text{ and } a + b \in \Delta, \forall x \in V,$$

and $\phi(x,a) = x$ for all $x \in V$ implies $a = 0$.

Because
$\Pi(0) = I_M$, where I denotes the identity transformation, we have

a) $\Pi(0,\underset{\sim}{x}) = \theta_o(\underset{\sim}{x}) = I(\underset{\sim}{x}) = \underset{\sim}{x}$.

b) $\Pi(\Pi(a,\underset{\sim}{x}),\ b) = \Pi_b(\Pi_a(x)) = \Pi(a + b,\ \underset{\sim}{x})$

$= \theta_{a+b}\ (\underset{\sim}{x})$

c) In a sufficiently small neighborhood of zero

$\Delta = [-\ \varepsilon,\ +\ \varepsilon] \in \mathbb{R}$ if for some $a \in \Delta$ it is true that $\Pi(a,x) = x$, for all $x \in \mathcal{M}$ then $a = 0$.

d) The set of all transformations Π_a forms a group. The group operation $\circ$ is composition of transformations. The situation does not change drastically if $\mathcal{M}$ is some subset of a Banach space instead of $\mathbb{R}^n$, and if the group $(G,\cdot)$ is a mapping from $\mathbb{R}^k$ into $\theta(M)$ instead of $\mathbb{R} \to \theta(\mathcal{M})$.

The requirements a), b), c), d) satisfied by a group G and the group representation $\Pi: \mathbb{R} \to \theta(\mathcal{M})$ (or $\mathbb{R}^k \to \theta(\mathcal{M})$) and the transformations $\theta(\mathcal{M})$ also form a differentiable manifold, then we have an example of a Lie group.

Definition. A Lie group is a differentiable manifold $\mathcal{M}$ with a group $(G,\circ)$ such that for any $\{x,y\} \in \mathcal{M}$ the group operation $(x \circ y^{-1}) = Z$ defines a C^∞ map $\{x,y\} \to Z$.

A historical comment. A somewhat simpler definition of a Lie group that did not require the C^∞ property of this map was originally offered in some earlier works.

2.2 Definitions of continuity and differentiability

For the sake of completeness we offer a brief review.

Let X, Y be Banach spaces. The map f: $\Omega \subset X \to Y$ is called strongly continuous in Ω if it is continuous with respect to the norm, that is, if for a sequence of points $\{x_i\} \in \Omega$

$$\lim \| x_i - x_o \| = 0 \text{ then } \lim \| f(x_i) - f(x_o) \| = 0.$$

A group of one parameter transformations $R \to \theta(\mathcal{M})$ is continuous at $a \in \mathbb{R}$ and the point $\tilde{x} \in \mathcal{M}$ if the map $\mathbb{R} \times \mathcal{M} \to \mathcal{M}$ is continuous at $\{a,\tilde{x}\}$.

Definitions of differentiability are almost identical. In each specific case the differentiability involves differentiability with respect to the parameter for a fixed $x \in \mathcal{M}$, and differentiability with respect to x for a fixed value of the parameter. Depending on the space X, the choice of a manifold $\mathcal{M}$ we may require Fréchet differentiability or Gateaux differentiability. If all requirements of differentiability are satisfied for a Group G we shall denote it by writing a superscript 1, i.e. $\mathbf{G}^1$.

2.3 Some common one-parameter groups

Some physically motivated examples include the following:

A1) The Galileo group

We consider a one-parameter group of transformations of the three-dimensional Euclidean space $\mathbb{R}^3$ of ordered triples $\{x,v,t\}$, $\pi((x,v,t),a) = (x',v',t')$ defined as follows: $t' = t$, $x' = x + at$, $v' = v + a$.

This is a solution of the initial value problem

$$\frac{dt'}{da} = 0 \quad , \qquad \frac{dx'}{da} = t \quad , \qquad \frac{dv'}{da} = 1.$$

We observe that the axiom (a) postulates

$\pi(x,v,t;0) = (x,v,t)$. We shall make without clarification the comment that laws of Newtonian mechanics are invariant under the action of the Galileo group.

A.2. The group of contraction (expansion) of $\mathbb{R}^n$ given by

$\pi(\underset{\sim}{x}, a) = x'$, is defined by the formula

$x_i' = x_i \exp(a\xi_i)$, where $\underset{\sim}{\xi}$ is a given constant vector, with $i = 1,2,\ldots n$.

This group is isomorphic to the Galilean group.

A.3 The group of linear homeomorphisms of a topological space X. (This is the group denoted by GL(X)).

This is usually written in physics in the form $x' = \ell(a) \cdot \langle \underset{\sim}{x} \rangle$, where $\langle x \rangle$ denotes the preimage of $\underset{\sim}{x}'$ under a linear transformation, and a is the parameter. We require that $\ell(0)$ is the identity transformation and that ℓ is a differentiable function. A specific example is obtained by identifying X with $\mathbb{R}^3$, and

$$\ell(a) = \begin{bmatrix} \cos a & \sin a & 0 \\ -\sin a & \cos a & 0 \\ 0 & 0 & 1 \end{bmatrix},$$

which is the group of rotations of $\mathbb{R}^3$ about an axis and corresponds to the coordinate change

$$\begin{cases} x^{1'} = x^1 \cos a + x^2 \sin a, \\ x^{2'} = -x^1 \sin a + x^2 \cos a, \\ x^{3'} = x^3. \end{cases}$$

A.4. <u>The Lorentz transformation</u>

$$x' = (x + vt)\,(1 - v^2/c^2)^{-\frac{1}{2}}$$

$$t' = t + \left(\frac{v}{c^2}\right) x \cdot (1 - v^2/c^2)^{-\frac{1}{2}}$$

$$y' = y$$

$$z' = z\ .$$

This can be simplified to a form

$x' = ct \cdot \sinh(a)$

$$ct' = ct \cdot \cosh(a),$$

$$y' = y,$$

$$z' = z,$$

where $\tanh a = v/c$.
The corresponding linear homeomorphism is given by

$$\ell(a) = \begin{bmatrix} \cosh a & \sinh a & 0 & 0 \\ \sinh a & \cosh a & 0 & 0 \\ 0 & 0 & 1 & 0 \\ 0 & 0 & 0 & 1 \end{bmatrix}$$

If v is very small compared to c, or if $v/c \to 0$ then the Lorentz transformation approaches the Galileo transformation. Thus, the Galileo group can be regarded as a limiting case of the Lorentz group.

A.5. The group $GL(n, \mathbb{C})$.

Let $\mathbb{C}^n$ denote an n-dimensional vector space over the complex numbers, with a basis $\{\mathbf{e}_i\}$, $i = 1,2,\ldots n$.

This could be regarded as a 2n-dimensional vector space over the real number field, with basis $\mathbf{e}_1, i\mathbf{e}_1, \mathbf{e}_2, i\mathbf{e}_2 \ldots, i\mathbf{e}_n$. The invertible linear transformations over $\mathbb{C}^n$ form a group denoted by $GL(n, \mathbb{C}) \cong GL(2n, \mathbb{R})$, where $\cong$ signifies isomorphism.

Example. Consider all linear transformations $\mathbb{R}^2 \to \mathbb{R}^2$ defined by the matrices $\begin{bmatrix} \alpha, & \beta \\ -\beta & \alpha \end{bmatrix}$, with $\alpha^2 + \beta^2 \neq 0$. All such transformations form a group $GL(2,\mathbb{R}) \cong GL(1, \mathbb{C})$ isomorphic to the group of non-zero complex numbers with ordinary multiplication as the group operation.

A.6. The subgroup $U(1)$ of the group $GL(1, \mathbb{C})$

This is a subgroup of $GL(1, \mathbb{C})$ of unimodular transformations. Regarding $GL(1, \mathbb{C})$ as the group of non-zero complex numbers under multiplication, $U(1)$ becomes the group of complex numbers having absolute value one, under multiplication. This is also known as "the circle group".

A.7. The group of transformations of $\mathbb{R}^3$ represented by 3 x 3 orthogonal matrices, usually denoted by O(3) and a subgroup of O(3) of the orthogonal matrices having the determinant equal to one, called a special orthogonal group SO(3).

A.8. The group L of linear fractional transformations of the complex plane plus a single point "at infinity" :

$$z' = \frac{az + b}{cz + d} \quad .$$

1.5 The tangent space for a one parameter group of transformations G^1.(the superscript one denotes that the group G is differentiable).

Let $\Delta \subset \mathbb{R}$, and
$\pi: \quad X \times \Delta \to Y$

be a group of one-parameter transformations of a metric space X into a metric space Y. Let us consider a specific (fixed) element $\underset{\sim}{x} \in X$. Then $\underset{\sim}{x} \times \Delta$ defines a parametrized curve in Y, which is called the orbit of the point x. That is, the orbit of $\underset{\sim}{x}$ is the curve $\pi(\underset{\sim}{x},a) = \underset{\sim}{x}' \in Y$. A more common notation is $\underset{\sim}{x}' = \pi_x(a)$. The corresponding derivative to the orbit of zero is the tangent vector space of the group G^1, i.e.

$$\underset{\sim}{\xi} : X \to Y, \ \underset{\sim}{\xi} = \underset{\sim}{\xi}(\underset{\sim}{x})$$

is given by

$$\underset{\sim}{\xi}(\underset{\sim}{x}) = \partial_a (\pi(\underset{\sim}{x},0)) .$$

Comment. The relation

$$\pi(\pi(\underset{\sim}{x},a),b) = \pi(\underset{\sim}{x}, a+b)$$

implies that

$$\partial_a \pi(\pi(\underset{\sim}{x},a),0) = \partial_a \pi(\underset{\sim}{x},a)$$

$$\partial_b \; \pi(\underset{\sim}{x},a),\; b=0) = \xi\cdot \pi(\underset{\sim}{x},a)$$

$$\partial_a \; \pi(\underset{\sim}{x},a) = \xi(0)\cdot\pi(x,a).$$

Therefore the group of transformations π solves the Cauchy problem for the system of differential equations

$$\partial_a \underset{\sim}{x}' = \xi(\underset{\sim}{x}'), \quad \underset{\sim}{x}'(0) = \underset{\sim}{x}.$$

This is the Lie equation associated with the Lie group π.

The following facts are well-known. The introduction of the tangent space is equivalent to local lineralization of the original problem after imbedding it in a larger one-parameter family of problems. This fact permits us in turn to apply group theoretic techniques to investigate properties of solutions of partial and ordinary differential equations. However, the problems related to ordinary differential equations reduce to finding the answer to the following question: Given a vector field ξ find a Lie group π for which ξ is the tangent space.

2.4 The infinitestimal generators of a group of transformations.

The infinitesimal generator acting on a mapping $F: X \to Y$ is given by the formula

$$(\xi\cdot\partial)\; F(x) = \partial F(x) < \xi \; (x) > =$$

$$\partial_a \; (F \circ \pi(x,a))\Big|_{a = 0}$$

If X is an n-dimensional space,

$$\underset{\sim}{x} = (x^1, x^2, .. x^n), \; \underset{\sim}{\xi} = (\xi^1, \ldots \xi^n) \text{ is a given}$$

constant vector, and

$$\partial = ({}^{\partial}/\partial x^1,\; {}^{\partial}/\partial x^2 \ldots\; {}^{\partial}/\partial x^n),$$

then $(\xi\cdot\partial) \equiv \xi^i \cdot {}^{\partial}/\partial x^i$.

Theorem. The mapping $F: X \to Y$ is an invariant of a group G^1 (i.e. $F(\pi(x,a)) = F(x)$)

if (for any $x \in X$) $(\xi \cdot \partial)\, F\, (x) = 0$.

The proof follows from the following lemma:

$\partial_a\, F(x') = (\xi \cdot \partial)\, F\, (x')$ for any differential map $F: X \to Y$.

Definition. An invariant $J: X \to Y$ of the group G^1 is called a universal invariant of G^1 if for any Banach space Z, and for any invariant $F: X \to Z$, there exists a map $\Phi: Y \to Z$ such that $F = \Phi \circ J$.

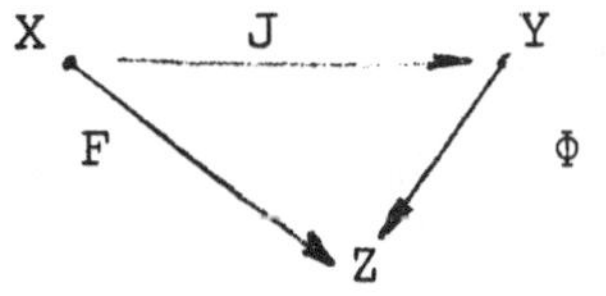

The system $$\xi^i \cdot \frac{\partial}{\partial x^i} F = 0$$

represents a total derivative if and only if

$$\frac{dx^1}{\xi^1} = \frac{dx^2}{\xi^2} = \cdots \frac{dx^n}{\xi^n} \quad . \tag{4.1}$$

The system arises in formulating the characteristic equations of a partial differential equation, and in considering the so called characteristic strips (see F. John [1]).

Examples of invariants

1) The group of translations of the space. The equation

$$(\xi \cdot \partial)\ F = 0$$

becomes

$$(x_0^i \cdot \partial/\partial x_i)\ F = 0.$$

2) The group of linear transformations of space. The invariants F must satisfy the condition

$$(a^i_j\, x^j \cdot \partial/\partial x_i)\; F = 0.$$

3) The group of rotations of $\mathbb{R}^3$. Let x^3 be the axis of rotation.
The invariance conditions are

$$(\xi\cdot\partial)\; F = 0 \qquad (-x^2 \frac{\partial}{\partial x_1} + x^1 \frac{\partial}{\partial x_2})\; F = 0.$$

2.5 The system $\dot{\tilde{x}} = \phi(a,\tilde{x})$.

2a. Basic group theoretic arguments. Note: The theory outlined below is known and can be found in Ovsiannikov's treatise [3].
Let us consider the first order system of ordinary differential equations:

$$(5.1)\quad \begin{cases} \dot{\tilde{x}} = \tilde{\phi}(a,x) & \tilde{x} = \{x_1, x_2 \ldots x_n\}, \\ \tilde{x}(0) = \tilde{x}_0, & \tilde{a} = \{a_1, a_2 \ldots a_m\} \text{ are} \end{cases}$$

parameters of the system

$$(\cdot = \frac{d}{dt}),$$

for which we seek a solution

$$(5.2)\quad \begin{cases} \tilde{\chi}_t = \tilde{\chi}(x_0,t) \text{ in } H^1_0\,[0,T], \\ \tilde{\chi}(\tilde{x}_0,0) = \tilde{x}_0, \\ \tilde{\chi}(\tilde{\chi}(\tilde{x}_0,t_1),t_2) = \tilde{\chi}(\tilde{x}_0,t_1+t_2), \end{cases}$$

where

i.e. we assume the semigroup property of solutions with respect to the (one) parameter t, keeping $\tilde{a}$= constant. We could do better if a prior knowledge of the system permits us to assume group theoretic property of solutions. Specifically, we assume that the mapping τ: $\tilde{\chi} \to \tilde{x}_t$ is invertible for all $t \in [0,T]$. The infinitesimal generator of the semigroup or group

τ_t: $t \to \underset{\sim}{x}(\underset{\sim}{x}_0, t)$ is

$$\left.\frac{d\,\underset{\sim}{x}(\underset{\sim}{x}_0,t)}{dt}\right|_{t=0} = \left.\underset{\sim}{\phi}(a,x(t))\right|_{t=0} .$$

The infinitesimal operator

$$(5.3) \quad Y = \sum_{i=1}^{n} \phi^i(\underset{\sim}{a},\underset{\sim}{x}) \cdot \frac{\partial}{\partial x^i}, \text{ such that}$$

$$(5.3^{a}) \quad Yz = \sum_{i=1}^{n} \phi^i(a,x) \frac{\partial z}{\partial x^i} = \frac{dz}{dt}$$

is of basic importance in this study. To find an invariant of the system (5.1) it suffices to find a function $w(\underset{\sim}{x})$ such that $Yw = 0$. It is easy to see that $w \equiv$ constant along any trajectory of (5.1) in that case.

Examples.

Consider the equation

$$\dot{\underset{\sim}{x}} = \underset{\sim}{c} ,$$

which has the solution

$$x_i = x_{o_i} + c_i t \quad , \qquad i = 1,2,\ldots n.$$

The corresponding group is the translation group in the direction of the vector

$$\{c_1, c_2, \ldots c_n\}^T .$$

The infinitesimal operator Y is given by

$$(5.4) \; Y = \sum_{i=1}^{n} \left(c_i \frac{\partial}{\partial x_i} \right)$$

Any function of $c_j x_i - c_i x_j$ is an obvious invariant.

There are exactly n-independent invariants of the translation group.

The other well known group of motions is the rotation group about an axis, specifically a subgroup of SO(3) of rotations about the z-axis.

In $\mathbb{R}^3$ we consider the vector equation

$$\frac{d}{dt}\begin{bmatrix} x \\ y \\ z \end{bmatrix} = \begin{bmatrix} \sin t & \cos t & 0 \\ -\cos t & \sin t & 0 \\ 0 & 0 & 1 \end{bmatrix} \cdot \begin{bmatrix} x_o \\ y_o \\ z_o \end{bmatrix} .$$

The infinitesimal operator Y is given by

$$Y = -y \frac{\partial}{\partial x} + x \frac{\partial}{\partial y} .$$

Here any function of z only, or any function of (x+y) is an invariant. Various functions of these invariants are also invariants.

Example:

$$\begin{cases} \dot{\underset{\sim}{x}} = \underset{\sim}{x} \\ \underset{\sim}{x}(0) = \underset{\sim}{x}_o \end{cases}$$

has solution $\underset{\sim}{x} = e^t \underset{\sim}{x}_o$. The infinitesimal operator Y is

$$Y = (\underset{\sim}{x} \cdot \frac{\partial}{\partial \underset{\sim}{x}}) = \sum_{i=1}^{n} x^i \frac{\partial}{\partial x_i} .$$

Invariants, which can be easily computed are x_i / x_j. For example, in $\mathbb{R}^2$ we have

$$x \frac{\partial}{\partial x}(x/y) + y \frac{\partial}{\partial y}(y/x) + x \frac{\partial}{\partial x}(y/x) + y \frac{\partial}{\partial y}(x/y) = 0,$$

as required.

2.6 A Lie algebra of infinitesimal operators.

We form a commutator of infinitesimal operators Y_1, Y_2:

$$[Y_1, Y_2] = Y_1 Y_2 - Y_2 Y_1 \tag{6.1}$$

In a manner of speaking the commutator measures the non-commutativity of Y_1 and Y_2. In our definition

$$[Y_1, Y_2] = Y_1 Y_2 - Y_2 Y_1 = - \underset{\sim}{\phi}_{(2)} \frac{\partial}{\partial \underset{\sim}{x}} \left(\underset{\sim}{\phi}_{(1)} \frac{\partial}{\partial \underset{\sim}{x}}\right) +$$

$$\phi_{(1)} \frac{\partial}{\partial \underset{\sim}{x}} \left(\underset{\sim}{\phi}_{(2)} \frac{\partial}{\partial \underset{\sim}{x}}\right) = -(D\,\phi_{(1)}, \phi_{(2)}) \frac{\partial}{\partial \underset{\sim}{x}} + (D\phi_{(2)}, \phi_{(1)}) \frac{\partial}{\partial \underset{\sim}{x}}$$

This is true because second derivatives cancel each other.

That is,

$$\left(\phi_2 \phi_1 \frac{\partial^2}{\partial \underset{\sim}{x}\, \partial \underset{\sim}{x}} - \phi_1\, \phi_2 \frac{\partial^2}{\partial \underset{\sim}{x}\, \partial \underset{\sim}{x}}\right)$$

is the zero operator when it is applied to the class of smooth functions. Here $D\,\underset{\sim}{\phi}$ is the is the Jacobian matrix.

For example, $\frac{D\,\underset{\sim}{\phi}}{D\underset{\sim}{x}}$ is given in $\mathbb{R}^3$ by the matrix

$$\begin{bmatrix} \frac{\partial \phi_1}{\partial x_1} & \frac{\partial \phi_1}{\partial x_2} & \frac{\partial \phi_1}{\partial x_3} \\ \frac{\partial \phi_2}{\partial x_1} & \frac{\partial \phi_2}{\partial x_2} & \frac{\partial \phi_2}{\partial x_3} \\ \frac{\partial \phi_3}{\partial x_1} & \frac{\partial \phi_3}{\partial x_2} & \frac{\partial \phi_3}{\partial x_3} \end{bmatrix}$$

Example.

Let

$$Y_1 = \frac{\partial}{\partial x} + y\frac{\partial}{\partial y}$$

$$Y_2 = \frac{\partial}{\partial x} + \frac{\partial}{\partial y}, \text{ then } [Y_1, Y_2] = -\partial/\partial y.$$

Definition.

The operators $Y_1, Y_2, \ldots, Y_k$ are called linearly independent in $\Omega \subset R^n$ if given $\hat{x} \in \Omega$ there do not exist constants $c_1(\hat{\tilde{x}}), c_2(\hat{\tilde{x}}), \ldots c_k(\hat{\tilde{x}})$, not all equal to zero, such that

$\sum_{i=1}^{k} c_i Y_i$ is the zero operator for each (fixed) value of $\hat{\tilde{x}} \in \Omega$. (This is the usual definition.)

<u>Lemma</u> 6.1

Let $\{Y_i\}$, $i = 1,2,\ldots, m \leq n$, **be** linear infinitesimal operators of the form (5.3) acting on a set of vectors $\sigma \subset R^n$. They are independent if and only if the rank of the $n \times m$ matrix

$$\left[\phi^i_{(\ell)}(x)\right], \quad \begin{array}{l} i = 1,2,\ldots,n \\ \ell = 1,2,\ldots,m \end{array}$$

(of the corresponding functions $\phi_{(\ell)}(a,\tilde{x}), x \in \sigma$, $(\ell = 1,2,\ldots,m)$ is exactly m.

Definition

The set of linearly independent operators $\{Y_k\}$ is <u>complete</u> if for any pair Y_i, Y_j the commutator $[Y_i,Y_j]$ is a linear combination of $\{Y_k\}$.

Lemma 6.2 If the system $\{Y_k\}$, $k = 1,2,\ldots,m$, $m < n$, is complete then the system

$$(6.1) \qquad \sum_{i=1}^{n} \phi^{i}_{(k)} \frac{\partial z}{\partial x^{i}} = 0, \qquad k = 1,2,\ldots,m,$$

has exactly n-m linearly independent solutions $\hat{z}_1, \hat{z}_2, \ldots \hat{z}_{n-m}$, which are invariants of motion for the system (5.1).

A complete system of linearly independent operators $\{Y_k\}$, $k = 1,2,\ldots, m < n$ forms the basis for a subalgebra of the Lie algebra of operators defined by the commutator operation.

Example.

$$\text{Let} \quad y_o = \frac{\partial}{\partial t} + \frac{\partial}{\partial x} + \frac{\partial}{\partial y} \quad ,$$

$$y_1 = \frac{\partial}{\partial t} + x \frac{\partial}{\partial x} \quad .$$

Then

$$[y_o, y_1] = \frac{\partial}{\partial x} \quad .$$

Denoting $y_2 = \frac{\partial}{\partial x}$ we obtain a system of three linearly independent operators y_o, y_1, y_2.

We could conclude that the group determined only by the operators y_o and y_1 does not have any invariants of motion.

Effects of changes in coefficients. We consider a dynamic autonomous system

$$\dot{\underset{\sim}{x}} = \phi(a, \underset{\sim}{x})$$

$$\underset{\sim}{x}(o) = \underset{\sim}{x}_o \quad ,$$

as in (5.1).

The infinitesimal operator Y_o is given by

$$(6.2) \qquad Y_o = \sum_{i=1} \phi^{i} (\underset{\sim}{a}, \underset{\sim}{x}) \frac{\partial}{\partial x^{i}}$$

We emphasize that Y_o is a function of $\underset{\sim}{a}$. Let us consider a functional $\phi(\underset{\sim}{a},\underset{\sim}{x})$.

A necessary condition for the invariance of Φ along any solution of (5.1) is:

$$(5.1) \qquad \sum_{i=1}^{n} \phi^{i}(a,\underset{\sim}{x}) \frac{\partial\phi(a,\underset{\sim}{x})}{\partial x^{i}} = 0.$$

Hence, if $x = \hat{x}(t)$ is the solution such that $\phi(\tilde{a},x) = \max_{a\varepsilon A} \phi(a,x) = \text{constant}$, and in the admissible region A, $\phi(x,a)$ is a differentiable function of a, then either

$$\frac{\partial\Phi}{\partial a}(\hat{x}(t),a)\Big|_{a=\hat{a}} = 0$$

or else $\hat{a}$ lies on the boundary of admissible region.

Examples of this straightforward approach are known and corresponding numerical techniques and examples can be found in [4], [5], [6] in the study of sensitivity of systems represented by ordinary differential equations.For systems of partial differential equations and for more abstract systems the equation (3.3) may serve as a starting point for a discussion of intimate relations between the group theoretic approach to the theory of invariants and the functional theoretic approach of M.M.Vainberg,M.Krein and the subsequent duality theory,as outlined in [7],[8].

For partial differential equations the determination of infinitesimal generators of the group of of transformations G^1 and finding of the corresponding invariants is generally quite difficult.

For example,let us consider a nonlinear wave equation

$$(6.4) \qquad w_{tt} = (f(a(x),w_x)w_x)_x.$$

One commences with specifying an infinitesimal generator of a group G^1 of the form

$$Y = \xi\frac{\partial}{\partial x} + \tau\frac{\partial}{\partial t} + \zeta\frac{\partial}{\partial w}$$

$$\xi = \xi(x,t,w),\ \tau = \tau(x,t,w),\ \zeta = \zeta(x,t,w)$$

and the extension of G^1 to

$$Y = Y + \alpha\frac{\partial}{\partial p} + \beta\frac{\partial}{\partial q} + \rho\frac{\partial}{\partial r} + \sigma\frac{\partial}{\partial s} + \gamma\frac{\partial}{\partial m},$$

where $p = \frac{\partial w}{\partial x},\quad q = \frac{\partial w}{\partial t},\ r = \frac{\partial^2 w}{\partial x^2},\ s = \frac{\partial^2 w}{\partial x \partial t},\ m = \frac{\partial^2 w}{\partial y^2}$

$$\alpha = (\frac{\partial}{\partial x} + p\frac{\partial}{\partial w})\zeta - p(\frac{\partial}{\partial x} + p\frac{\partial}{\partial w})\xi - q(\frac{\partial}{\partial x} + p\frac{\partial}{\partial w})\tau, \text{ etc.}$$

Specifically, the equation

$$(6.5^a) \quad W_{tt} = \frac{1}{2}(W_x^2)_x$$

has the group of admissible infinitesimal operators that is generated by the basis of the Lie algebra given by

(6.6)

$$Y_1 = \frac{\partial}{\partial x}, Y_2 = \frac{\partial}{\partial t}, Y_3 = \frac{\partial}{\partial w}, Y_4 = t\frac{\partial}{\partial w}, Y_5 = t\frac{\partial}{\partial t} - 2w\frac{\partial}{\partial w}$$

$$Y_6 = x\frac{\partial}{\partial x} + 3w\frac{\partial}{\partial w}.$$

Linear forms of these operators determine G^1 and allow us to find invariants defined by this group. These results are known and could be found in the fundamental work of Ovsiannikov[3] and his collaborators. See for example [6], [21] [28] and [32].

7. Variation of the domain.

7.1 Change of shape.

We introduce a one parameter family of maps.

Let us suppose that a shape $\Omega \subset \mathbb{R}^n$ is dynamically deformed with the shape Ω_t uniquely defined for all values of $0 \leq t \leq 1$ and the map $T_t: \Omega \to \Omega_t$ is a continuous homeomorphism. Each point $\underset{\sim}{x} \in \Omega$ is continuously moving along a simple arc $T_t: x \to x_t$, where $T_{t=0} = I$, is the identify map on Ω. We presume that T_t defines a strongly continuous family of operators which form a semigroup under composition:

$$T_\mu \circ T_t (\Omega) = T_\mu (\Omega_t) = T (\Omega_{\mu+t}) = T_{\mu+t}(\Omega).$$

The infinitesimal generator of this semigroup is given by

$$(7.1) \qquad \lim_{t\to 0} \frac{T_t - I}{t} = \dot{T}_0 \, .$$

$\dot{T}_0$ defines physically the initial velocity operator. Specifically,

(7.2)

$$\dot{T}_0 x_0 = \lim_{t\to 0} \frac{1}{t} [T_t(x_0) - x_0] = \lim_{t\to 0} \frac{1}{t} [x_t - x_0] = \left.\frac{dx_t}{dt}\right|_{t=0}$$

In a sufficiently small neighborhood of zero one can estimate the deformation at x_0 by writing

$$(7.3) \qquad x_t = x_0 + t \cdot \dot{T}_0 x_0 + r(t, x_0),$$

where $r(t, x_0)$ is the remainder obeying the limit relation:

$$(7.3^{\underline{a}}) \qquad \lim_{t\to 0} \frac{1}{t} r(t, x_0) = 0 \, .$$

The function $\dot{T}_0 x_0$ defines the sensitivity of the shape Ω to the deformation process described by the operator family T_t computed at $x_0 \in \Omega$.

For an arbitrary function, or functional $\Phi(x)$, $x \in \Omega$, $\Phi: \Omega \to \mathbb{R}$ we define the material (Lie) derivative along the action of the semi-group of transformation T_t to be

(7.4)
$$\lim_{\tau \to 0} \frac{\Phi_\tau(x+\tau\dot{T}_0 x) - \phi(x)}{\tau} = \dot{\Phi}(x),$$

where $\Phi_\tau(y) = \phi(y_\tau)$.

Similarly for an arbitrary function Z

(7.5)
$$\dot{Z} = \lim_{\tau \to 0} \frac{1}{\tau} \cdot \{Z_t(x + t\,\dot{T}_0 x) - Z(x)\}$$

In problems involving continuum mechanics or other continuous phenomena pointwise definitions are inappropriate and our definition should be corrected to read $\dot{Z}(x)$ is defined almost everywhere in the $H_0^m(\Omega)$ sense by the relation

(7.5a)
$$\lim_{\tau \to 0} \left\| \frac{1}{\tau} \cdot \{ Z(x+t\dot{T}_0 x) - Z(x)\} - \dot{Z}(x) \right\|_{H_0^m(\Omega)} = 0.$$

$\dot{Z}$ is frequently called the material derivative.

Here $H_0^m(\Omega)$ is the appropriate Sobolëv space assigned to our problem. By the Sobolëv imbedding lemma, if $2m > n$, $H_0^m(\Omega)$ is a subspace of $C(\Omega)$ (the class of continuous functions) and pointwise definitions make sense. Otherwise, only $L^2(\Omega)$ averages make physical and mathematical sense and all concepts defined above must be interpreted in the sense of $H_0^m(\Omega)$ averaged quantities. As the shape of Ω is transformed, the basic state equations

(7.6)
$$\begin{cases} Az = f & \text{in } \Omega, \\ z \equiv 0 & \text{on } \partial\Omega \end{cases}$$

are conserved. That is

(7.6$\underline{a}$)
$$A_{z_t} = f \qquad \text{in } \Omega_t,$$
$$z_t = 0 \qquad \text{on } \partial\Omega_t.$$

The system (7.6$\underline{a}$) could be interpreted as a weak equation in $H_0^m(\Omega)$. The variational form of equation (7.6$\underline{a}$) is obtained in Ω_t:

(7.7)
$$a_t(z_t,\bar{z}_t) = \ell_t(\bar{z}_t) \qquad \text{in } \Omega_t$$
$$z_t \in \mathcal{Z}_t,$$

where

(7.8)
$$a_t(z_t,\bar{z}_t) = (A_{z_t}, \bar{z}_t)_{\Omega_t},$$

(7.9)
$$\ell_t(\bar{z}_t) = (f,\bar{z}_t)_{\Omega_t}.$$

The linear form $a(z_t,\bar{z}_t)$ is regarded as the

Friedrichs form, that is, both the domain and range of the operator A has been changed. Hence, it is not the same operator but the extension of A. However, we use the same symbol for the operator A as before, since no confusion can arise.

7.2. The material derivative.

Let J be the Jacobian matrix associated with the transformation T_τ.

(7.2.1)
$$\begin{cases} J_\tau = \dfrac{\partial T_\tau}{\partial x} = I + \tau\dfrac{\partial \dot{T}_0}{\partial x} + r(\tau^2), & \left(\dfrac{\partial T}{\partial x} = \dfrac{\partial T_{\tau_i}}{\partial x_j}\right) \\ J_{\tau=0} = I. & \text{(the identity operator)} \end{cases}$$

By assumption the map T_t is a homeomorphism, and the matrix J_τ is nonsingular for values of τ considered here.
Hence J_τ^{-1} exists.

A simple computation shows that

(7.2.2$\underline{a}$)
$$\lim_{\tau \to 0} |J_\tau| = \operatorname{div} (\dot{T}_0),$$

and taking the limit $0 = \lim_{\tau \to 0} \frac{d}{d\tau} \left| JJ^{-1} \right|$, one obtains

(7.2.2$\underline{b}$)
$$\lim_{\tau \to 0} \left| J_\tau^{-1} \right| = - \operatorname{div} (\dot{T}_0).$$

We define the material derivative of a functional $\phi(f_\tau) = \int_{\Omega_\tau} f_\tau(x_\tau)\, d\,\Omega_\tau$ computed at $\tau = 0$ to be the functional:

(7.2.3)
$$\phi_0' = \frac{d}{d\tau} \int_\Omega [f_\tau \,(x+\tau \dot{T}_0) \cdot J\,]\, d\Omega \Big|_{=0}$$

$$= \int_\Omega [f'(x) + (\nabla f(x), \dot{T}_0) + (f(x) \operatorname{div} \dot{T}_0)]\, d\Omega$$

$$\int_\Omega [f'(x) + \operatorname{div} (\dot{T}_0 f)]\, d\Omega \quad ,$$

where f' denotes the partial derivative

$$f'(x) = \lim_{\tau \to 0} \frac{f_\tau(x) - f(x)}{\tau} \quad .$$

Using the divergence theorem, one can transform equation (7.2.3) to the form

$$\Phi_0' = \int_{\Omega} f'(x)\, d\Omega + \int_{\partial\Omega} f(x)(\dot{T}_0 \cdot \underset{\sim}{n})\, d(\partial\Omega). \tag{7.2.4}$$

here $\underset{\sim}{n}$ is the unit vector normal to $\partial\Omega$.

Sufficient regularity conditions have to be assumed concerning $\partial\Omega$ for the second (boundary) integral to make sense.

The discussion of smoothness of the domain is beyond the scope of this book. In fact the dependence of fairly complex functionals, which occur in the engineering practice on specific shape of the domain has not been satisfactorily treated in the existing literature, and the most important questions associated with this problem are not yet answered. The results of J. Cea and his collaborators such as [10], [11] continue an important idea of J. Hadamard [12] but so far have not progressed beyond analysis of certain numerically motivated developments. The inclusion of domain parameters complicates the basic theory and makes it difficult to go beyond fairly heuristic approaches such as described in [16] or in the monograph of N.V. Banichuk [13]. Once the domain is regarded as fixed and only design parmeters related to energy functionals are discussed, one can use the functional analytic techniques as given in [4], [5] and follow well established numerical techniques, such as the gradient projection method.

We comment that certain forms and expressions given in this work are difficult to interpret from this point of view or to identify with physical concepts or with purely mechnical quantities. Other forms fit naturally into both the functional analytic and physical interpretation. For example, the generators of certain Lie groups, as given above, are rather difficult to analyze, except if one regards these operators in the context of a mapping which preserves the finiteness of energy in an appropriate Sobolëv space setting.

However, $a(Z,\lambda)$ may be identified with a bilinear form naturally arising in physical considerations, such as, for example, the virtual work performed on a structural system by external loads.

7.3 <u>An Application to design practice.</u>

We analyze the beam design optimizing the total weight. The cost functional Φ is given by

$$\Phi = \int_0^{\ell} \rho\, A(x)\, dx.$$

The material derivative of Φ is given by

(7.3.1)

$$\Phi' = \int_0^{\ell} (\rho A)'\, dx + \int_{\partial\Omega} (\rho A)(\dot{T}_0 \cdot \underset{\sim}{n})\, ds$$

$$= \int_0^{\ell} (\rho A)'\, dx + \rho A\dot{T}_0(\ell) - \rho A\dot{T}_0(0).$$

$$= \rho \frac{d}{d\tau} \int_0^{\ell} (A_\tau + (x\ \tau\dot{T}_0) \cdot J_\tau)\, dx\Big|_{\tau=0} + \rho\, A\dot{T}_0(\ell),$$

since we can vary the length by keeping one end fixed, with no loss of generality, Φ' cannot be determined unless we first establish the mapping, or the class of admissible mappings $T_t: \Omega \to \Omega_t$.

Such a mapping is easily established by fairly "conventional" (i.e. by now 5 - year old, or older techniques) following either direct methods such as given in [16] or more sophisticated methods introduced in [17] and further expanded by Haug and his associates in a series of articles. (See for example [18], [19]).

To illustrate this point we offer a fairly easy computation of sensitivity of the compliance functional for an elastic beam.

We assume the Euler-Bernoulli linear beam theory with a distributed load $q(x) \in L_2\, [0,\ell\,]$ and the specific weight of the beam $g\rho A(x)$ contributing to the applied load.

The natural setting for this problem is the Sobolëv space in which the inner product is

$$<f,g> = \int_{\Omega} (EI(x) f_{xx} g_{xx}) dx.$$

We write the basic equation of equilibrium for an elastic beam

$$(7.3.1) \quad (EI(x)W_{xx})_{xx} = q(x) + g\rho A(x),$$

equating the second derivative of the bending moment with the load applied (q(x)) plus the weight of the beam regarded as additional loading. We assume constant geometry and a relation $I(x) = \phi(A(x))$, which is associated with certain additional geometric assumptions of our model. $\phi(A(x))$ is assumed to be a differentiable function of A(x).

As usual, ρ is the material density; E - the Young modulus, A(x) - the cross-sectional area, I(x) - the moment of inertia of the cross-sectional area about the neutral axis, g is the earth's gravitational constant.
The potential energy binomial form is given by

$$(7.3.2) \quad a(w,\lambda) - b(w,\lambda) =$$

$$\int_0^{\ell} [E\, \phi(A(x))\, W_{xx}\, \lambda_{xx}]\, dx -$$

$$\int_0^{\ell} (q(x)\lambda + g\, \rho\, A(x)\, \lambda)\, dx.$$

The load q(x) may depend on w, and for the sake of greater generality we shall assume that

$$q(x) = \bar{q}\, (w(x), x).$$

Let the cost functional be given by the compliance

$$\Phi(w) = \int_0^{\ell} [\big(q(w(x),x) + \rho\, g\, A(x)\big) w(x)]\, dx$$

The corresponding bilinear form is given by

$$\hat{\Phi}(w,\lambda) = \int_0^{\ell} [\,(q(w(x),x) + \rho g A(x))\cdot \lambda(x)]\, dx \quad .$$

The change of shape of the beam consists of changing the parameter A(x) and the length ℓ. The Fréchet derivative with respect to A(x) is computed directly:

$$\delta_A(b(w,\lambda)) = \{{}_0\int^{\ell}(q_A\lambda + g\rho)\, dx\}\ \delta_A$$

$$\delta_A(a(w,\lambda)) = \{{}_0\int^{\ell}[\,(E\phi'(A)\ W_{xx}\ \lambda_{xx})dx$$

$${}_0\int^{\ell}[\,(E\phi(A)\ W_{A_{xx}}\ \lambda_{xx})]dx\ \}\ \delta A,$$

where δA is an arbitrary admissible variation of the area function A(x).
Hence,

$$\delta_A(V) = \delta_A\ (a(w,\lambda) - b(w,\lambda))$$

$$= \{{}_0\int^{\ell}[\,(E\phi'\ (A)\ W_{xx}\lambda_{xx}) + (E\phi(A)(W_A)_{xx}\ \lambda_{xx})]\ dx\}\cdot \delta A(x)$$

$$- \ \{\ {}_0\int^{\ell}(q_A\lambda + g\rho)\ dx\}\cdot \delta A$$

The variation of the total weight is given by the material derivative

$$[\int_0^{\ell} \rho\ A(x)\ dx]' = \rho A\dot{T}\Big|_0^{\ell}$$

$$= \rho A(\ell)\dot{T}(\ell) - \rho A(0)\dot{T}(0).$$

The material derivative of the potential energy is given by

$$(7.3.3)\quad V' = \int_0^{\ell} [\,(EI(A)\ \lambda_{xx})_{xx}\ (\lambda_x\dot{T}_t) - (\underline{q}\ \lambda_x\dot{T}_t)]\ dx$$

$$+ \ EI(A)\lambda_{xx}\ (\lambda_x\dot{T}_t)_x\Big|_0^{\ell} - (EI(A)\lambda_{xx})_x$$

$$(\lambda_x \dot{T}_t)\Big|_0^{\ell} + q(x)\lambda - EI(A)(\lambda_{xx})^2\, \dot{T}_t\Big|_0^{\ell}$$

Identifying λ and W and simplifying this expression by assuming the beam to be clamped - clamped, one obtains an extremely simple sensitivity formula in terms of the boundary transformation rate

$$V' = -EI(A)\, W_{xx}^2\, \dot{T}_\tau\Big|_0^{\ell} .$$

This conclusion is fairly obvious if some intuitive physical arguments are offered. However, this is not the case if one considers even simple cases of plate or shell design.

More challenging numerical computations using this technique are also given in the monograph of Haug, Choi and Komkov [17] 1986, Academic Press, New York).

However, even such simple one-dimensional analysis may offer nontrivial insight into the design alteration procedures.

Example. We consider one of the extensively researched class of problems in structural mechanics - that of optimization, sensitivity and differentiability of the natural frequencies. If we accept the Euler - Bernoulli model for a vibrating beam, the sensitivity of a simple eigenvalue corresponding to the fundamental natural frequency is given by

$$(7.3.4) \qquad \zeta' = -2 \int_0^{\ell} \{ EI(A(x))W_{xx}(W_x\dot{T}_t)_{xx} + \zeta\,\rho\, A(x)\, W\, (W_x\dot{T}_t)\}\, dx + [EI(A(x))(W_{xx})^2 - \zeta\rho A(x)W^2]\dot{T}_t\Big|_0^{\ell}$$

If we assume the clamped-clamped support conditions at $x = 0$ and $x = \ell$ and recall that $W(x)$ is an eigenvector corresponding to the simple eigenvalue ζ we can derive the simplified sensitivity formula

$$\zeta' = - EI(A(x))(W_{xx})^2 \dot{T}_t \Big|_0^\ell$$

indicating the effects in changes of natural frequency with changes in length and in the bending moment $EIW_{xx} = M(x)$ and W_{xx} at end points. It is clear that ζ' is a quadratic function of the "symmetric moment":

$\mathfrak{M}(x) = \sqrt{EI} \cdot W_{xx}$ at the end points of the beam. To decrease the natural frequency one should move outward the end point of the beam at which $|\sqrt{EI} \cdot W_{xx}|$ is bigger.

The complex problem of "crossing of eigenvalues" resulting from an iterative application of such optimizing procedure (for constant total-weight) is deliberately avoided here.

8.0 The Noether Approach

In the celebrated paper [20] of 1918 Emmy Noether established an algorithm for deriving invariants and laws of conservation for the physical systems modeled by a system of differential equations. Specifically the Noether paper [20] utilized the properties of the Lagrangian action integral to derive some identities for the solutions of corresponding Euler-Lagrange variational equations. The groups of transformation that arise in a general statement of Noether's theorem are called Noether's groups. Their generalizations, as given for example,in [21], are the Lie-Bäcklund groups of transformations.

The main topic of this chapter are the Noether groups and applications of Noether's theory to engineering design with only limited applications of the Lie-Bäcklund theory.

As in classical problems of the calculus of

variations we consider extremization of a functional $J(\tilde{x})$ over some class of admissible functions $\tilde{x}(t)$. However, in engineering problems the formulation of the mathematical model depends on the design considerations, on identification of parameters, on the choice of admissible forcing terms, and on admissibility of certain classes of functions representing physical quantities.

In the original formulation of Noether's theory (see [20]) the fundamental functional, that is the functional that is to be minimized, is the action integral of Lagrange and de Maupertuis. Let us consider a typical, but simple structural example.

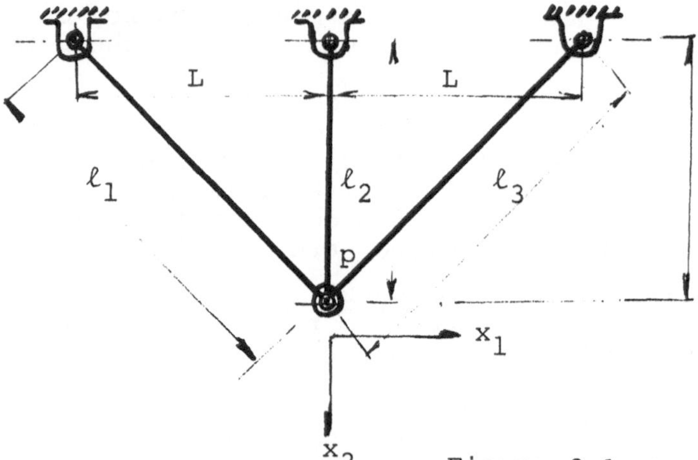

Figure 8.1

We assume linear elasticity, Hooke's law, and equate the potential energy with the strain energy. We require invariance of the potential energy of the system under coordinate changes. The trigonometric relations for undeflected structure are:

$$\ell_1 = \ell_3 = \sqrt{2}\ \ell_2 = \sqrt{2} \cdot L\ .$$

Deflection of the point p is given by the vector $\tilde{x} = [x_1, x_2]$ and is supposedly small. The forces in the bars labelled 1,2,3 are respectively

$$P_1 = A_1 E \frac{\delta \ell_1}{\ell_1},\quad P_2 = A_2\, E \frac{\delta \ell_2}{\ell_2},\quad P_3 = A_3\, E \frac{\delta \ell_3}{\ell_3},$$

where $\delta\ell_1 = \frac{\sqrt{2}}{2}\ (x_1 + x_2)$

$$\delta\ell_2 = x_2$$

$$\delta\ell_3 = \frac{\sqrt{2}}{2}\,(x_2 - x_1)\,.$$

If $\ell_1 = \ell_3$, $A_1 = A_2 = A_3$,

then the work done by the forces P_1, P_2, P_3 is equated with the strain energy and is given by:

$$(8.1)\quad U = \frac{AE}{4L}\left\{ \frac{(x_1 + x_2)^2}{\sqrt{z}} + 2x_2^2 + \frac{(x_2 - x_1)^2}{\sqrt{z}} \right\}$$

This can be rewritten in a convenient quadratic form

$$U = \langle \underset{\sim}{x},\ K\underset{\sim}{x}\rangle =$$

$$(8.2)\quad [x_1,\ x_2] \begin{bmatrix} \dfrac{AE}{2\sqrt{2}\ L} & 0 \\ 0 & \dfrac{AE}{2L}\left(\dfrac{1}{\sqrt{2}} + 1\right) \end{bmatrix} \begin{bmatrix} x_1 \\ x_2 \end{bmatrix}$$

The matrix K is the stiffness matrix for this truss.
The equation (8.2) is of the form $J = \langle x,\ Ax\rangle$

where A is a linear operator acting in an appropriate vector space, which in this case is a two-dimensional Euclidean space.
Note: for reason of convenience the engineering texts prefer to write $U = \frac{1}{2}\langle x,\ \underset{\sim}{K}x\rangle$. This is a trivial matter, requiring no special attention.

It appears that our choice of coordinates $[x_1,\ x_2]$ for the "state of the system" $\underset{\sim}{x}$ was particularly advantageous. The matrix A is symmetric and the strain energy contains only quadratic terms in $\{x_1, x_2\}$. Written out in full the equation (8.2) is $U = \sum_{i,j=1}^{2} a_{ij}\, x_i\, x_j$ with all $a_{ii} \geq 0$, and

$$\sum_{i,j} a_{ij}\, x_i\, x_i > c \sum_i x_i^2$$

for some $c > 0$.

All these conditions are trivially satisfied here. Moreover, in our case all $a_{ij} = 0$ if $i \neq j$, But let us suppose that we were less fortunate, or less smart and that we have chosen some other coordinate system $\{\xi_1, \xi_2\} = \underset{\sim}{\xi}$ that is describing "the state" of the two-degrees of freedom three bar truss as show on figure 8.1 Let $\underset{\sim}{x}$ be given in terms of coordinate vector $\underset{\sim}{\xi}$.

$\underset{\sim}{x} = Q\cdot\underset{\sim}{\xi}$, where Q is a 2×2 invertible matrix. Then

$$U = \langle \underset{\sim}{x}, K\underset{\sim}{x} \rangle =$$

$$\langle Q\,\underset{\sim}{\xi}, K\,Q\,\underset{\sim}{\xi} \rangle = \tag{8.3}$$

$$\langle \underset{\sim}{\xi}, Q^*\,K\,Q\,\underset{\sim}{\xi} \rangle,$$

where Q^* is in this case the transpose of Q. An analogue of (8.3) is applicable to any quadratic form. Let <,> denote an inner product in some Hilbert space H.

If $J = \langle \underset{\sim}{x}, A\,\underset{\sim}{x} \rangle$, $\underset{\sim}{x} \in H$, $A : H \to H$, is a linear operator, and $\underset{\sim}{\xi} = Q\,\underset{\sim}{x} \in H$, Q is a linear transformation mapping basis $\underset{\sim}{x}$ of H into basis $\underset{\sim}{\xi}$ of H, then

$$J = \langle \underset{\sim}{x}, A\,\underset{\sim}{x} \rangle = \langle \underset{\sim}{\xi}, Q^*\,A\,Q\,\underset{\sim}{\xi} \rangle,$$

where Q^* is the adjoint of Q.

The coordinate changes appear to be intimately related to the preservation of the strain energy of the system. No major theoretical difficulties arise in this computation.

However, the problem posed here was not the "real life" problem facing an engineer who wishes to design a three bar truss. In our formulation the configuration is given, the truss is already designed. The question has not been even asked, whether this is a "good" truss or not. Also, the problem was modeled by a simplified discrete version called "lumped parameters" where each member of the truss was of uniform shape and had a constant cross-section. Kinetic energy was ignored. Changes in the design were not discussed. Should we consider members having variable cross-section, and consider the kinetic energy effects, then the potential energy is replaced in all known considerations by the Lagrangian action integral.

In a simple Hookean model the Lagrangian energy density is given by

$$(8.4) \qquad L(\underset{\sim}{x}) = \tfrac{1}{2} \sum_{j=1}^{3} \left\{ \int_0^{\ell_j} [\rho A_j(x)](\dot{x})^2 \, dx - \int_0^{\ell_j} \frac{P_j^2 \, dx}{EA(x)} \right\},$$

$$\sum_{j=1}^{3} P_{jx_1} = \mathbf{P}_{x_1}, \quad \sum_{j=1}^{3} P_{jx_2} = \mathbf{P}_{x_2},$$

where the subscripts x_1 and x_2 denote the x_1 and x_2 components of the force vectors $\mathbf{P}_j$, $j=1,2,3$, or of the total force $\mathbf{P}$ applied to the junction point p. The improvement in the design for the three bar truss subjected to a single load $\mathbf{P}$ is a fairly simple procedure if we assume constant cross-sections and lump all properties.

Let us suppose σ_0 is the known yield stress of the material. Let $p_y = p_{x_2}$ denote the vertical displacement of the point p. The strains are given by $\varepsilon_i = \frac{\Delta\ell_i}{\ell_i}$, $i = 1,2,3$, where $\Delta\ell_1$ (and $\Delta\ell_3$) is found from the trigonometric relations

$$\frac{P_y}{\sin\theta} = \frac{\ell_1}{\sin(\frac{\Pi}{4} - \theta)} = \frac{\Delta\ell_1 + \ell_1}{\sqrt{2}/2}. \text{ while } \Delta\ell_2 = P_{x_2} = P_y$$

The forces corresponding to the yield stress are $P_1 = \sigma_o A_1 = P_3 = \sigma_o A_3$, $P_2 = \sigma_o A_2$, where A_1, A_2, A_3 are, respectively, cross-sectional areas of members 1,2,3. Also we have

$P_1/\sqrt{2} + P_3/\sqrt{2} + P_2 = P$. Thus we have a very simple linear programming problem to determine

$\max_{\{A_i\}} P$ with the constraint $A_1\ell + A_2\ell_2 + A_3\ell_3 =$ constant $= \hat{C}$. That is, we seek values of A_1, A_2, A_3 which give max P, subject to

$P_i \leq \sigma_o A_i$, and to some minimum size restrictive condition $A_i > A_{o_i}$, with $A_{o_i} < \frac{1}{3} C/\ell_2$, $i = 1,2,3$.

The optimization problem is not quite that simple if we consider the "distributed parameter" case with the Lagrangian action integral

$\int_{t=0}^{T} L \, dt$ with L given by the relation (8.4).

One possible approach may utilize the theory of invariants. A well-established technique has been developed in theoretical physics using a group theoretic approach developed originally by Emmy Noether ([20] , 1918).

8.1 Invariance of the action integral

Let the Lagrangian density function be

$L = L(t, \underset{\sim}{x}(t), \dot{\underset{\sim}{x}}(t))$, $\underset{\sim}{x} = (x_1, x_2 \ldots x_n)$ and let the corresponding action integral be

$$J(\tilde{x}) = \int_0^T L(t,\tilde{x}(t), \dot{\tilde{x}}(t))\, dt.$$

We introduce a one parameter group of transformations G^1:

$$t \to \bar{t}$$

$$\tilde{x} \to \bar{\tilde{x}} ,$$

$$\bar{t} = \phi^o(t, \tilde{x}, \varepsilon), \ \bar{x}^i = \phi^i(t, \tilde{x}, \varepsilon), \ i = 1,2,\ldots n.$$

that such

$$(8.5) \quad \begin{cases} \phi_o(t,\tilde{x}, 0) = t \\ \phi_i(t,\tilde{x}, 0) = x_i \ . \end{cases}$$

The functions ϕ_i, $i=0,1,2,\ldots n$, are smooth mappings from a domain $\Omega \subset \mathbb{R}^n \times I$ into I ($I \subset \mathbb{R}$), with Ω- an open connected region of $\mathbb{R}^{n+1}$, containing the origin.

The linear terms of the Taylor series in powers of ε are called the infinitesimal generators of the group G^1. These are

$$\xi^i\ (t,\tilde{x}) = \lim_{\varepsilon\to 0} \frac{\partial\phi^i(t,\tilde{x},\varepsilon)}{\partial\varepsilon} = \frac{\partial\phi^i(t,\tilde{x},0)}{\partial\varepsilon} ,$$

$i = 0, 1,2,\ldots n.$

For convenience we could denote $\xi^o(t,x) = \tau$ because of the mental association of $\phi^o(t,\tilde{x})$ with a different time scale τ.

The fundamental result of Noether concerns the properties of the groups of transformations $(t,\tilde{x}) \to (\tau,\tilde{\xi})$ for which the action integral is an invariant.

Physical considerations require certain energy forms to be independent of any (arbitrary) choices of coordinates. Specifically, the action integral assumes a minimal value that should be described by a specific number that depends only on the chosen units of energy but should be completely independent of the choice of any coordinate system.

Instead of "full invariance" we could substitute local invariance i.e. invariance for small values of the parameter ε, or some other definition of approximate invariance. Noether considered linear or "absolute invariance", when the action integral is invariant provided that terms containing powers of ε of order two and higher are neglected.

Let $J(x) = \int_0^T L(t, \underset{\sim}{x}(t), \dot{\underset{\sim}{x}}(t))\, dt$,

and
$$\bar{J}(\xi,\varepsilon) = \int_0^{\bar{T}} \bar{L}\left(\tau, \underset{\sim}{\xi}(\tau), \frac{d\xi}{d\tau}; \varepsilon\right) d\tau.$$

Then

$$(8.6)\quad J(\underset{\sim}{x}) - \bar{J}(\underset{\sim}{\xi},\varepsilon) = o(\underset{\sim}{\varepsilon}),$$

or

$$(8.6^a)\quad \lim_{|\varepsilon|\to 0} \frac{L(t,\underset{\sim}{x}(t),\dot{x}(t)) - \bar{L}(\tau,\xi,\frac{d\xi}{dt}, \varepsilon)\frac{d\tau}{dt}}{|\varepsilon|} = 0$$

$$\text{i.e. } \Delta = L - \bar{L}\,\frac{d\tau}{dt} = o(\underset{\sim}{\varepsilon})$$

Note: We have used a one-parameter group of transformations only for the sake of clarity of presentation. No basic changes are required if ε is replaced by $\underset{\sim}{\varepsilon} = (\varepsilon^1, \varepsilon^2 \ldots \varepsilon^k)$, that is if we consider a k-parameter family of coordinate transformations. The infinitesimal generators of the corresponding group G^1 are

$$(8.7)\quad \begin{cases} \tau_j = \dfrac{\partial\phi^o(t,\underset{\sim}{x},0)}{\partial\varepsilon^j} \\ \xi^i_j = \dfrac{\partial\phi^i(t,\underset{\sim}{x},0)}{\partial\varepsilon^j} \end{cases}$$

The divergence invariance, or invariance up to a divergence term is defined for the action integral

(8.8) $J(x) = \int_0^T L(t,x,\dot{x})$, if

$$\Delta = \bar{L}\,(\tau,\underset{\sim}{\xi}(\tau),\dot{\underset{\sim}{\xi}}(\tau),\underset{\sim}{\varepsilon})\frac{d\tau}{dt} - L(t,\underset{\sim}{x}(t),\underset{\sim}{x}(t))$$

$$= \sum_{j=1}^{k} \varepsilon^j \frac{d\tilde{\Phi}_j}{dt} + o(\varepsilon), \text{ for some functions}$$

$$\tilde{\Phi}_j(t,\underset{\sim}{x}(t),\underset{\sim}{\varepsilon}) \in C^1\,(\Omega).$$

The divergence invariance becomes absolute invariance if $\Delta = o(\varepsilon)$ and full invariance if $\Delta \equiv 0$. (See 8.6a)

The Noether invariance theorem states that a necessary condition for the invariance up to a divergence term of the action integral

$$\int_0^T L(t,\underset{\sim}{x}(t),\ \dot{\underset{\sim}{x}}\ (t)\ )\ dt\ ,$$

under the action of a k-parameter group of transformations Π (where the parameter vector $\underset{\sim}{\varepsilon} = (\varepsilon_1,\ldots\varepsilon_k)$, and $\Pi\colon\ (t,\underset{\sim}{x}) \to (\tau,\ \underset{\sim}{\xi},\ \underset{\sim}{\varepsilon})$) is the existence of k-functions

$\tilde{\Phi}_j(t,\underset{\sim}{x}) \in C^1(\Omega)$, $j = 1,2,\ldots k$, such that the

following equations are true.

$$(8.9)\qquad [L - \dot{x}^i(\frac{\partial L}{\partial \dot{x}^i})]\,\tau_j + \frac{\partial L}{\partial \dot{x}^i}\,\xi^i_j - \Phi_j = \text{constant}$$

(see, for example Logan [37] for the derivation.)

An example.

Consider the equation

a) $\dfrac{\partial^2 u}{\partial r^2} + \dfrac{1}{r}\dfrac{\partial u}{\partial r} + k^2u^3 = 0,$

where k is a real constant, with the associated Lagrange integral

$$(*)\quad J = \int_{r=o}^{R} r\,[\tfrac{1}{2}(u')^2 - \frac{k^2}{4} u^4]\,dr = \int_{o}^{R} L(r,u,u')dr$$

We seek a one-parameter group of transformations of the form

$$\left.\begin{aligned} \bar{r} &= r + \varepsilon\,\rho \\ \bar{u} &= u + \varepsilon\,\nu \end{aligned}\right\},$$

$\rho=\rho(r,u)$, $\nu=\nu(r,u)$.

Keeping the Lagrangian integral invariant, the Noether invariance formula is

$$\frac{\partial L}{\partial r}\rho + \frac{\partial L}{\partial u}\nu + \frac{\partial L}{\partial u'}\left(\frac{\partial \nu}{\partial r} - u'\frac{\partial \rho}{\partial r}\right) + L\frac{d\rho}{dr} = 0,$$

where $\frac{d\rho}{dr} = \frac{\partial \rho}{\partial r} + \frac{\partial \rho}{\partial u}u'$.

Written in full the Noether formula is

$$(\tfrac{1}{2}(u')^2 - \frac{k^2}{4}u^4)\rho - (\frac{k^2}{3}ru^3)\cdot\nu + (ru')\cdot(\frac{\partial \nu}{\partial r} - u'\frac{\partial \rho}{\partial r})$$

$$+ r\,[\frac{(u')^2}{2} - \frac{k^2}{4}u^4]\,(\frac{\partial \rho}{\partial r} + \frac{\partial \rho}{\partial u}u') = 0.$$

We equate separately to zero terms containing different powers of u'.

$$\left.\begin{aligned} &(i) \quad r\frac{\partial \rho}{\partial r}u + \rho u + \frac{4}{3}r\nu = 0 \\ &(ii) \quad -\frac{k^2}{4}u^4\frac{\partial \rho}{\partial u} + \frac{\partial \nu}{\partial r} = 0 \\ &(iii) \quad \tfrac{1}{2}\rho - \tfrac{1}{2}r\frac{\partial \rho}{\partial r} = 0 \\ &(iv) \quad \frac{\partial \rho}{\partial u} = 0 \end{aligned}\right\} \quad (**)$$

Thus, from (iv) we conclude that $\rho = \rho(r)$
Equation (iii) becomes

$$\rho - r\frac{d\rho}{dr} = 0.$$

It has a simple solution
$\rho = r$ (or $\rho = Cr$ where C is an arbitrary constant).
Also, from (ii) we derive $\frac{\partial \nu}{\partial r} = 0$

and $\nu = \nu(u)$, and the relation (i) becomes

$\nu = -\frac{3}{2}u$. Thus the Lagrangian integral (*) is

invariant under the transformation

$$\bar{r} = r(1 + \varepsilon)$$

$$\bar{u} = u(1 - \frac{3}{2}\varepsilon)$$

This is only one (particular) transformation derived by regarding k as a constant and by equating separately each term of the power series in u' to zero.

Example. Invariance of the Lagrangian action integral under time translation.
We introduce a one-parameter group of transformations

$$\bar{t} = t + \varepsilon$$
$$\bar{x}^i = x^i$$

The infinitesimal generators are

$$\tau = 1$$
$$\xi^i = 0, \quad i = 1, 2, 3,$$

and all remaining infinitesimal generators are all zero.
The first integral of motion is

$(L - \dot{x}^i \frac{\partial L}{\partial \dot{x}^i})\,\tau + \frac{\partial L}{\partial \dot{x}^i}\,\xi^i$ = constant, which is

$$L - \dot{x}^i \frac{\partial L}{\partial \dot{x}^i} = L - \dot{x}^i p_i = \text{constant} .$$

But

$$-(L - \dot{x}^i p_i) = H$$ is the Hamiltonian. Thus, the Lagrangian integral is invariant under time translation if and only if the Hamiltonian is contant along any trajectory of motion.

An example.

We consider an elastic inextensible beam in bending modeled by a chain consisting of two links. We simplify the dynamic model by considering it to be a double pendulum with torsion springs modeling elastic resistance against bending

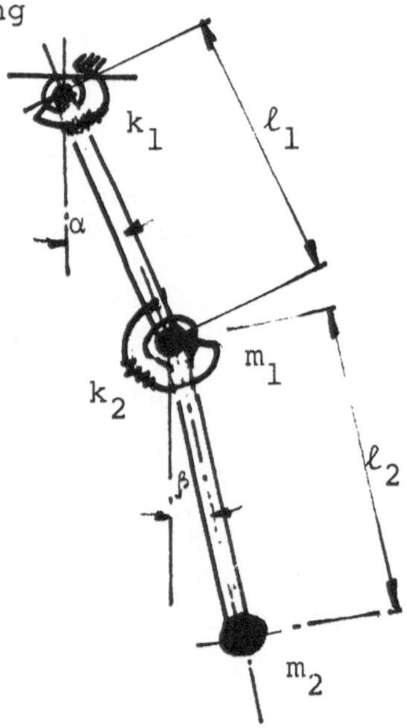

Figure 1.1

The kinetic energy is

$$T = \tfrac{1}{2}(m_1 + m_2)\, \ell_1^2\, \dot{\alpha}^2 + \tfrac{1}{2}\, m_2\, \ell_2^2\, \dot{\beta}^2 + m_2(\ell_1 \ell_2\, \dot{\alpha}\, \dot{\beta}) \cos(\alpha - \beta)$$

and the potential energy is

$$V = -(m_1 + m_2)\, g\, \ell_1 \cos\alpha - m_2 g\, \ell_2 \cos\beta$$

$$+ \frac{1}{2} k_1 \alpha^2 + \frac{1}{2} k_2 (\beta - \alpha)^2$$

The action integral is $\int_{t=o}^{t=T} (T - V)\, dt = \int_o^T L\, dt$.

The equations of free motion are

$$(m_1+m_2)\, \ell_1^2\, \ddot{\alpha} + m_2\, \ell_1\ell_2 \cos(\alpha-\beta)\, \ddot{\alpha} - m_2\, \ell_1\ell_2\, \dot{\alpha}\dot{\beta}$$
$$\sin(\alpha-\beta) - (m_1+m_2)\, g\ell_1 \sin\alpha + k_1\, \alpha_1 + k_2(\alpha-\beta) = 0,$$
$$m_2\ell_2^2\, \ddot{\beta} + m_2\, \ell_1\ell_2 \cos(\beta-\alpha)\ddot{\beta} - m_2\, \ell_1\ell_2\, \dot{\alpha}\dot{\beta} \sin(\beta-\alpha)$$
$$+ g\, m_2\, \ell_2 \sin\beta - k_2(\beta-\alpha) = 0.$$

It is not hard to show that these equations are invariant under time translation and so is the action integral.
Let us take

$$\bar{\alpha} = \alpha$$

$$\bar{\beta} = \beta$$

$$\bar{t} = t + \varepsilon.$$

The one parameter group (which is a subgroup of the Galilean group) has all infinitesimal generators equal to zero except $\partial\bar{t}/\partial\varepsilon\,(\varepsilon = 0) = 1$ and, trivially, the action integral is divergence invariant.

The Noether necessary condition is

$$\frac{\partial L}{\partial \tau} \cdot \dot{\tau} = \frac{\partial L}{\partial t} = \frac{d\Phi}{dt} ,$$

where $\varepsilon \frac{d\Phi}{dt} = L(\alpha,\beta,t) - L(\bar{\alpha},\bar{\beta},\bar{t})$

$+ \mathcal{O}(\varepsilon) = L(\alpha,\beta,t) - L(\alpha,\beta, t+\varepsilon) + \mathcal{O}(\varepsilon).$

Since $\frac{\partial L}{\partial t} = \frac{dL}{dt}$ (See Part 1, Chapter 1, page 73),
the divergence invariance is proved.

As an exercise the reader could try to prove invariance under simple (angle) translation

$$\bar{\alpha} = \alpha + \varepsilon_1$$

$$\bar{\beta} = \beta + \varepsilon_2$$

$$\bar{t} = t \quad .$$

At this time we should point out that α,β are the "natural" coordinates in this two-degrees of freedom mechanical system. The design parameters k_1, k_2, m_1, m_2, ℓ_1, ℓ_2 are regarded as constants. This is not a realistic treatment of the "real" engineering problems in which such double pendulum models a behavior of a component of some fairly complex mechanism or structure. For example, the behavior of a long gun tube subjected to recoil forces and reactions of the mount may be modeled by a finite element multiple pendulum(with n-degrees of freedom) similar to the one illustrated on figure 1.1. The cost functional could be either the maximum deflection during one vibration cycle, or the numerical value of the lowest (but non-zero!) natural frequency.

In dealing with such optimization problems we must consider not only the changes of the"natural" coordinates α , β and of the time scale, but also changes of the design parameters.

Thus,the equations of motion (*) are now regarded as a system of relations in a higher dimensional space in which "coordinates" α,β,m_1, m_2, ℓ_1, ℓ_2, k_1, k_2 all can be varied according to Noether's rule.

9. Conservation of kinetic energy integral and the de Maupertuis principle.

If external forces are absent, de Maupertuis principle asserts that the actual motion of a mechanical system minimizes the integral

$$(9.1) \quad \int_{t_0}^{T} T(\underset{\sim}{x}, \dot{\underset{\sim}{x}}, t)\, dt \quad , \underset{\sim}{x} = (x_1, x_2, \ldots x_n)$$

Let us suppose that the kinetic energy is of the form

$$T(x, \dot{x}, t) = \tfrac{1}{2}\sum_{i,j} g_{ij}\, \dot{x}^i \dot{x}^j \ ,(\text{where}$$

$$g_{ij} = g_{ij}(\underset{\sim}{x}) \text{ is the kinematic metric})$$

and that t does not explicitely occur in the kinetic energy form, i.e. $\frac{\partial T}{\partial t} = 0$.

Let us consider a one-parameter transformation

$$\left.\begin{aligned} \bar{t} &= t \\ \bar{x}^i &= x^i + \varepsilon\, \xi^i(x) \end{aligned}\right\}$$

(9.1)The Noether formula for invariance of the integral (9.1) after some simplifications is

$$\tfrac{1}{2}\left(g_{ik}\frac{\partial \xi^k}{\partial x^j} + g_{kj}\frac{\partial \xi^k}{\partial x^i}\right)\dot{x}^i \dot{x}^j = 0$$

In this content $\dot{x}^i$ are to be chosen arbitrarily. Thus,the necessary condition for the invariance of (9.1) is

$$(9.1^a) \quad g_{ik}\frac{\partial \xi^k}{\partial x^j} + g_{kj}\frac{\partial \xi^k}{\partial x^i} = 0 \quad , \quad i,j = 1,2,\ldots n.$$

These are the Killing equations that are fundamental in differential geometry and particle mechanics. Their solutions are not easy to find, are not necessarily unique, and may not even exist at all.

9.2 An example: Optimal design of a column.

Introduction:

Using Noether's technique we derive some invariants associated with the problems of optimization of design. While specifically a beam design problem is discussed here, the technique can be generalized in an obvious manner to thin plate, or shell problems. The specific problem discussed in this part concerns the optimization of a column. Some invariants derived here as examples of the technique can be used to point out the basic difficulties discovered by much harder arguments in many recent papers. In particular, our method points out the existence of singularities which are associated with an optimal column design even in the absence of multiple eigenvalues in the Euler's eigenvalue problem, which corresponds to the buckling phenomenon.

In the present example we use Noether's ideas to derive various integral invariants associated with different classes of optimization problems for elastic systems. Our goal posed in this example is to derive physically significant invariants which are directly related to the problem of optimization of elastic beams. We shall consider in particular the problems of optimization of the cross-sectional area of a column against buckling, and a closely related problem of optimization of a vibrating beam.

Let us consider the linear (approximate) differential equation representing Euler's buckling problem

9.2 $$(EI(x)\,y'')'' + P_{cr}y'' = 0\,, \qquad y(0) = y(l) = 0\,,$$

which can be derived from the energy equality

9.3 $$\tfrac{1}{2}\int_0^l (EIy'')^2\,dx = \tfrac{1}{2}\int_0^l P_{cr}(y')^2\,dx\,.$$

We regard, however, the cross-sectional area of the beam as a variable A = A(x), but assume the existence of a functional relation $I(x) = \tilde{I}(A(x))$ between the moment of inertia of the cross-sectional area I(x) and its magnitude A(x). A well explored class of optimization problems consists of either maximizing the first eigenvalue P_{cr_1} subject to the constraint

$$W = \int_0^l A(x)\,dx = \text{constant}\,, \tag{9.4}$$

or of minimizing W subject to the constraint $P_{cr_1} \geqq K > 0$, where K is a given value. See [13] [14],[22],[23] for a sampling of relevant literature. Reference [23] is of particular importance since it is largely responsible for many current trends. The beam optimization problem consists in choosing $A(x) \geqq 0,\ 0 \leqq x \leqq l$, such that

$$\min_{y \in Y} \left\{ \int_0^l E\tilde{I}(A(x))\,(y'')^2\,dx \Big/ \int_0^l (y')^2\,dx \right\} \geqq K\,, \tag{9.5^a}$$

and

$$\int_0^l A(x)\,dx = \text{minimum}\,. \tag{9.5^b}$$

The class of admissible displacement functions Y consists of $H_0^2[0,1]$ functions which satisfy all assigned boundary conditions at x = 0 and at $x = \ell$, such as

$$y(0) = y(l) = 0\,, \qquad I(0)\,y'' = I(l)\,y''(l) = 0$$

(a free support at each end).

We postulate that the solution A(x) must be a non-negative continuous function whose derivative dA/dx exists almost everywhere, such that

$$\int_0^l (EA(x))^{-1}\,dx < \infty\,. \tag{9.6}$$

We postulate $A(x) \in H_0^1[0\,\ell]$, and the condition (9.4) For the physical reasoning justifying the necessity of the constraint (9.6) see [24]

The minimization problem as stated above in (9.5) is associated with stationary behavior of the functional

$$\lambda(A, y) = \mu_0 \int_0^l A(x)\,\mathrm{d}x - \frac{1}{2}\beta\left(\int_0^l E\tilde{I}(A(x))\,(y'')^2\,\mathrm{d}x - K\right) - \frac{\mu}{2}\left(\int_0^l (y')^2\,\mathrm{d}x - 1\right)$$

$$= \int_0^l \lambda(0)\,\mathrm{d}x\,, \qquad (9.7)$$

where β, μ are Lagrangian multipliers. More specifically, if $\lambda(A)$ is a Fréchet differentiable functional of the vector $A \varepsilon H_1[0,\ell]$, then $\partial\lambda/\partial A=0$ is a necessary condition for the optimality of $\lambda : H^1 \to \mathbb{R}$, and the vanishing of $\partial\lambda/\partial y$ is in fact the statement that the equation

$$(EIy'')'' + P_{cr}y'' = 0$$

is satisfied, combined with the inequality

$$P_{cr} \geqq K$$

(which, by the way, can be replaced by $P_{cr} = K$, without any loss of generality).

Following the classical formulation of Emmy Noether, we introduce a k-parameter $(\varepsilon_1, \varepsilon_2, \ldots, \varepsilon_k)$ continuous group of transformations G_k:

(*) $\quad \bar{x} = \varphi(A, x, y, \varepsilon)\,, \qquad \bar{y} = \psi(A, x, y, \varepsilon)\,, \qquad \bar{A} = \nu(A, x, y, \varepsilon)\,,$

such that

(**) $\quad \bar{x} = \varphi(A, x, y, 0) = x\,, \qquad \bar{y} = \psi(A, x, y, 0) = y\,, \qquad \bar{A} = \nu(A, x, y, 0) = A\,.$

The transformations $(x, y, A) \to (\bar{x}, \bar{y}, \bar{A})$ do form a group and must have a uniquely defined inverse for each admissible value of the parameter vector ε. This presents a problem of choosing the class of transformations which is "naturally" suited to our physical problem. Since the Noether invariants depend on our choice of the transformation, the depth of the results is directly related to our ability of choosing the group G.

The infinitesimal generators of the group G are given by

$$\tau_i = \frac{\partial\varphi}{\partial\varepsilon_i}\bigg|_{\varepsilon=0}, \qquad \xi_i = \frac{\partial\psi}{\partial\varepsilon_i}\bigg|_{\varepsilon=0}, \qquad \zeta_i = \frac{\partial\nu}{\partial\varepsilon_i}\bigg|_{\varepsilon=0}, \qquad i = 1, 2, \ldots, k\,.$$

We shall choose some k-parameter groups G_k of transformations, mapping

$$x \in [0, l] \text{ into } \bar{x}, \qquad y \in H_0^2[0, l] \text{ into } \bar{y} \in H_0^2[0, l],$$
$$A > 0, \qquad A \in H_0^1[0, l] \text{ into } \bar{A} \in H_0^1[0, l].$$

We comment that in general $\bar{x}$, $\bar{y}$, $\bar{A}$ do not satisfy physical requirements imposed on x, y, and A. For example $\bar{A}$ may become negative, $\bar{y}(x)$ may exhibit unrealistic shapes,etc. This is immaterial, since the transformation applied to x,y,A serves only as a means for deriving some properties of the functional (9.7) and thereby of the whole class of buckling or vibration problems. We do not have to regard it as a physically admissible transformation. The functional λ is invariant under the action of the group G_k if for each i = 1,2,...,k, the equation (9.9)

$$\frac{\partial\lambda}{\partial x}\tau_i + \frac{\partial\lambda}{\partial A}\eta_i + \frac{\partial\lambda}{\partial y}\xi_i + \frac{\partial\lambda}{\partial y_x} + \left(\frac{d\xi_i}{dx} - y_x\frac{d\tau_i}{dx}\right) + \frac{\partial\lambda}{\partial y_{xx}}\left(\frac{d^2\xi_i}{dx^2} - 2y_{xx}\frac{d\tau_i}{dx} - y_x\frac{d^2\tau_i}{dx^2}\right) + \lambda\frac{d\tau_i}{dx} = 0$$

is satisfied.

Since in the case of beam optimization represented by functional (9.9) λ is not an explicit function of x or of y, the equation (9.9) may be simplified into the form

$$(9.8) \qquad \frac{\partial\lambda}{\partial A}\eta_i + \frac{\partial\lambda}{\partial y_x}\left(\frac{d\xi_i}{dx} - y_x\frac{d\tau_i}{dx}\right) + \frac{\partial\lambda}{\partial y_{xx}}\left(\frac{d^2\xi_i}{dx^2} - 2y_{xx}\frac{d\tau_i}{dx} - y_x\frac{d^2\tau_i}{dx^2}\right) + \lambda\frac{d\tau_i}{dx} = 0,$$

where d/dx denotes a total derivative; for example

$$\frac{d\xi_i}{dx} \equiv \frac{\partial\xi_i}{\partial x} + \frac{\partial\xi_i}{\partial y}y_x + \frac{\partial\xi_i}{\partial y_x}y_{xx} + \frac{\partial\xi_i}{\partial y_{xx}}y_{xxx} + \frac{\partial\xi_i}{\partial A}\frac{\partial A}{\partial x}.$$

In the classical application of Noether's theory, the transformation consisting of time translation gives the invariance of the Hamiltonian, if the Lagrangian does not implicitly depend on time. Regarding our functional λ as "the Lagrangian", and noting that λ does not explicitly depend on x, we obtain an invariant of the system by setting

$$\bar{x} = x + \varepsilon_1, \qquad \bar{A} = A, \qquad \bar{y} = y.$$

Then $\tau_1 = 1$ and the remaining infinitesimal generators of G: τ_i, ξ_i, η_i are all equal to zero. Hence, we have

$$\lambda - y_x \frac{\partial \lambda}{\partial y_x} + y_x \frac{d}{dx}\left(\frac{\partial \lambda}{\partial y_{xx}}\right) - y_{xx} \frac{\partial \lambda}{\partial y_{xx}} = \text{constant}\,, \qquad (9.10)$$

or if we use (9.7)

$$\left.\begin{aligned} &\frac{d}{dx}\left[\lambda + \mu(y')^2 - \beta y' \frac{d}{dx}(EIy'') + \beta EI(y'')^2\right] = 0\,, \\ \text{and} \quad & \\ &\frac{\partial \lambda}{\partial A} = 0\,. \end{aligned}\right\} \qquad (9.10^a)$$

These relations are valid for all $x \in [0, l]$. The term

$$\frac{\partial \lambda}{\partial y_x} - \frac{d}{dx}\left(\frac{\partial \lambda}{\partial y_{xx}}\right) = p \qquad (9.11)$$

is analogous to the canonical momentum for a Lagrangian which depends on both velocity and acceleration (See H.S.P. Grässer[25])

The expression (9.10) is analogous to the Hamiltonian

$$H = L - \dot{x}p - \ddot{x}\frac{\partial L}{\partial \ddot{x}}, \quad \text{where} \quad p = \frac{\partial L}{\partial \dot{x}} - \frac{d}{dt}\left(\frac{\partial L}{\partial \ddot{x}}\right). \qquad (9.12)$$

This Hamiltonian is constant along any trajectory of motion if the Lagrangian L does not explicitly depend on time. We observe also that λ does not explicitly depend on y(x).

Introducing the transformation

$$\bar{x} = x\,, \quad \bar{A} = A\,, \quad \bar{y} = \varepsilon + y\,,$$

we obtain the analogue of a well-known invariant of classical mechanics

$$\frac{\partial \lambda}{\partial y_x} - \frac{d}{dx}\left(\frac{\partial \lambda}{\partial y_{xx}}\right) = \text{constant}\,. \qquad (9.13)$$

Inserting (9.10) into (9. 7) we have

$$-\mu y' + \beta \frac{d}{dx}(EI(A)\,y'') = \text{constant}\,. \qquad (9.13^a)$$

Inserting this expression into (9.10) we obtain

$$\lambda + y'\,(\text{constant}) - y_{xx}\frac{\partial \lambda}{\partial y_{xx}} = \text{constant}\,, \qquad (9.13^b)$$

or for some constant α

$$\lambda + \alpha y' + \beta EI(y'')^2 = \text{constant}\,, \qquad (9.13^c)$$

and

$$\frac{\partial\lambda}{\partial A} = 0 , \qquad (9.13^d)$$

valid for all $x \in [0, l]$.

We are ready to experiment with more complex identities. We consider the transformation (*)-(**) of the form

$$\bar{x} = x + \varepsilon_1 x(l - x) , \qquad \bar{y} = x + \varepsilon_2 y , \qquad \bar{A} = A + \varepsilon_3 y + \varepsilon_4 A ,$$

restricting the values of x to the interval $0 \leq x \leq 1$, and $A \geq 0$. The mapping $(x,y,A) \to (\bar{x},\bar{y},\bar{A})$ is one to one and has an inverse, if x is restricted to the interval [0,1].

We compute the infinitesimal generators of the corresponding group G_4,

$$(9.14) \quad \tau_1 = x(l - x) , \quad \tau_2 = \tau_3 = \tau_4 = 0; \quad \xi_1 = 0 , \quad \xi_2 = y , \quad \xi_3 = \xi_4 = 0;$$
$$\nu_1 = \nu_2 = 0 , \quad \nu_3 = y , \quad \nu_4 = A .$$

The functional λ is invariant under this 4-parameter family of transformations if

$$(9.15) \quad \frac{\partial\lambda}{\partial x}\tau_i + \frac{\partial\lambda}{\partial A}\nu_i + \frac{\partial\lambda}{\partial y_x}\left(\frac{d\xi_i}{dx} - y_x\frac{d\tau_i}{dx}\right) + \frac{\partial\lambda}{\partial y}\xi_i + \frac{\partial\lambda}{\partial y_{xx}}\left(\frac{d^2\xi_i}{dx^2} - 2y_{xx}\frac{d\tau_i}{dx}\right.$$
$$\left. - y_x\frac{d^2\tau_i}{dx^2}\right) + \lambda\frac{d\tau_i}{dx} = 0 .$$

The system (9.14) after substituting the identities (9.15) becomes

$$\frac{\partial\lambda}{\partial x}x(l - x) - \frac{\partial\lambda}{\partial y_x}(y_x \cdot (l - 2x)) - \frac{\partial\lambda}{\partial y_{xx}}(y_{xx}(l - 2x) - 2y_x) + \lambda(l - 2x) = 0 , \qquad (9.16)$$

$$\frac{\partial\lambda}{\partial y_x}(y_x) + \frac{\partial\lambda}{\partial y_{xx}}(y_{xx}) + \frac{\partial\lambda}{\partial y}y = 0 , \quad \frac{\partial\lambda}{\partial A}y = 0 , \quad \frac{\partial\lambda}{\partial A}A = 0 . \qquad (9.17)$$

Equations (9.16 and (9.17) state that

$$(9.18) \qquad \frac{\partial\lambda}{\partial A} = 0 \quad ,$$

except at points where (y = 0 , A = 0). Observing that $\frac{\partial\lambda}{\partial x} \cdot \frac{\partial\lambda}{\partial y} = 0$, and substituting (9.18) into (9.17), we obtain the following system

$$(9.19) \qquad (y \cdot A)\frac{\partial\lambda}{\partial A} = 0 , \qquad \frac{\partial\lambda}{\partial x} = \frac{\partial\lambda}{\partial y} = 0 ,$$

$$\frac{\partial\lambda}{\partial y_x} y_x + \frac{\partial\lambda}{\partial y_{xx}} y_{xx} = 0\,, \qquad -2y_x \frac{\partial\lambda}{\partial y_{xx}} + (l-2x) = 0\,, \tag{9.20}$$

expressing the necessary conditions for the invariance of λ. Moreover the following "conservation laws" are immediately derived

Along the length of the column the following energy density forms are constant:

$$C_i = \lambda\tau_i + \left(\frac{\partial\lambda}{\partial y_x} - \frac{d}{dx}\left(\frac{\partial\lambda}{\partial y_{xx}}\right)\right)(\xi_i - y_x\tau_i) + \frac{\partial\lambda}{\partial y_{xx}}\frac{d}{dx}(\xi_i - y_x\tau_i)\,. \tag{9.21}$$

Writing these identities in detail, we obtain

$$C_1 = \lambda \cdot x(l-x) - \left(\frac{\partial\lambda}{\partial y_x} - \frac{d}{dx}\left(\frac{\partial\lambda}{\partial y_{xx}}\right)\right)(x(l-x)\,y_x) - \frac{\partial\lambda}{\partial y_{xx}}\frac{d}{dx}(x(l-x)\,y_x)\,, \tag{9.22^a}$$

$$C_2 = \left(\frac{\partial\lambda}{\partial y_x} - \frac{d}{dx}\left(\frac{\partial\lambda}{\partial y_{xx}}\right)\right) y + \frac{\partial\lambda}{\partial y_{xx}} y_x\,, \qquad C_3 = C_4 \equiv 0\,. \tag{9.22^b}$$

Substituting (9.6) into (9.22^b), we obtain

$$C_2 = \left(-\mu y_x + \beta\frac{d}{dx}(EIy_{xx})\right) y - \beta EIy_{xx}y_x = -\mu y y_x + \beta y\frac{d}{dx}(EI(x)\,y_{xx})$$

$$-\beta y_x EIy_{xx}\,, \tag{9.23}$$

which is a constant of the deformation. and

$$\frac{dC_2}{dx} \equiv 0\,, \qquad \forall x \in [0,l]\,.$$

A numerical procedure:

The required energy density function is given by

$$\lambda(x) = \mu_0 A(x) - \frac{1}{2}\beta EI(A(x))\,y_{xx}^2 - \frac{\mu}{2}y_x^2 - k\,, \tag{9.24}$$

where k is a constant. $k = \dfrac{\beta K + \mu}{2l}$.

Using the relations (9.13^a) and (9.13^c) we proceed as follows. The equation (9.24) states

$$\mu_0 = \frac{1}{2}\beta EI'(A)\,y_{xx}^2\,, \quad \left(' = \frac{d}{dA}\right). \tag{9.25}$$

For simplicity's sake let us assume I(A)=CA^2, so that I'(A) ≅ A, the constant factor being absorbed by β, and choose $\mu_0 \equiv 1$. Then

$$y_{xx}^2 = 2/(\beta EA)\,. \tag{9.26}$$

Using the invariant C_2 given by equation (9.23) we have

$$y_x(-\mu y - \beta EI y_{xx}) + \beta y \frac{d}{dx}(EI y_{xx}) = C_2 .$$

Substituting $I \sim A^2$ and using the equation (9.23) i.e.

$$-\mu y_x + \beta \frac{d}{dx}(EI y_{xx}) = C_3 \quad , \qquad (9.23^b)$$

we obtain

$$C_3 y - \beta E A^2 y_x y_{xx} = C_2 \qquad (9.23^d)$$

or the equation

$$y_x + kA^{-3/2} y = CA^{-3/2} \quad , \qquad (9.23^e)$$

where k and C are constants, with initial condition $y(0) = 0$.

We can solve this equation in a closed form

$$y(x) = \exp\left(-\int_0^x kA^{-3/2}(\xi)\, d\xi\right)\left\{\int_0^x \left(CA^{-3/2}(\xi)\left[\exp\left(\int_0^x kA^{-3/2}(\tau)\, d\tau\right)\right]\right) d\xi\right\}. \quad (9.27)$$

Because of the expected symmetry $\int_0^x$ can be replaced by $\int_\ell^{\ell-x}$ without altering the value of $|y(x)|$.

Hence, if $A(x)$ is known at any point $x \in [0,1]$, $y(x)$ can be calculated after a choice of constants k and C. If $y(x) \neq 0$, y_x can be found from the relation (9.23^e)

However, if $y(x)$ is known at three consecutive points of a grid imposed on the beam, then $A(x)$ can be calculated from the relation (9.23^e) This suggests an iterative scheme which would completely avoid the usual techniques used in optimization and rely on the use of integral invariants to implement the numerical procedure.

9.3. Comments

Since we are solving in a systematic manner a first-order differential equation (9.23^e) no boundary condition can be assigned to it, unless the solution has discontinuities in y_x, or else if by some extreme stroke of luck the boundary conditions assigned to $y(x)$, i.e. $y(0) = y(\ell) = 0$

just happen to be satisfied. Equation (9.25) clearly indicates that at such points of discontinuity of y_x the cross-sectional area A(x) must be equal to zero, that is a hinge develops at such singular points of the optimal design.
The author has come to the conclusion that while such singularities are mathematically justified, the fault lies with the mathematical description of the physical phenomena.

The most frequent occurence is lack of consistency between results predicted by a theory and (either explicit or inexplicit) assumptions that were made in formulating such theory. For example, in Volume 1 of this work we derive some results based on "slow deformation" and "quasi-static assumptions" which, roughly speaking presume that energy is almost conserved, as a structure deforms in reaching an equilibrium. That is, the work performed by outside loads is almost equal to the total potential energy of the structure as it attains an equilibrium. The trouble is that at an equilibrium point the total potential energy must be stationary, and for a stable equilibrium it must attain a minimum. As long as these two statements produce somewhat similar predictions concerning the state of the structure in a stable equilibrium, we have nothing to worry about. Otherwise, the assumptions of the theory must be reexamined and more realistic assumptions need to be incorporated into the mathematical model. A more common problem of incorrect modelling results from assumptions of "smallness" of deflections, angles, strains, etc. For example, in the Euler-Bernoulli theory of beams and in the Lagrange-Sophie Germaine theory of plates or in the Kirchhoff-Love theory we assume that something is small and second order terms of such small quantities can be safely disregarded. If using this theory we discover that in the predicted "states" these quantities are not small and their products are important, we need to reject all such predicted results, revise our assumption and start again using a more exact theory. This type of feedback between assumptions and predictions is too frequently not instituted even in cases involving stability or deflection of very slender structures, where doubts regarding correct theoretical modelling should be raised almost routinely.

Almost any practicing engineer has come across some computations of structural designs that were using "cookbook" formulas of the type $\sigma = Mc/I$ in cases where they were clearly not applicable, because some hypothesis of Euler-Bernoulli theory were violated by large rotations of cross-sectional areas, large linear and/or angular deflections, large inertia terms ignored in the static theory, or very high stress concentrations that locally invalidated the Hooke's law.

Similarly, computations using ideal gas equations (say $pv = GRT$) may have been "routinely" applied in a temperature range where at least the ionization and dissociation phenomena must be considered. The velocity of sound may have been derived from the "cookbook formula" $(g\gamma RT)^{1/2}$ to be used in equations of internal ballistics. (This particular horror may be discovered in some reports whose authors should request protection under the fifth amendment to the U.S. Constitution.) Linearity has been assumed in contradiction of existing data, and so on. This is not the place to expand the sound principles of mathematical modelling of physical phenomena, except that the breaking of symmetry phenomena discussed in this volume have been rather badly neglected by the engineering community just for such reasons. Only some recent theoretical advances allowed us to understand more clearly the corresponding modelling theories that have been so well adapted into modern physics. While abuses of the modelling principles are common, a remarkable progress has been made towards better comprehension by many authors such as C. Truesdell (and his followers), S. Antmann, J.T. Oden, E. Haug (and the Iowa School), P. Pedersen, F. Niordson, N. Olhoff (and the Danish School), A.P. Seyranian, N.V. Banichuk, Chernous'ko (Russian School), J. Cea, B. Rousselet (French School) raising the level of sophistication of modern modelling and engineering design theory.

10. Dual Form of Noether's Theorem with Applications to Continuum Mechanics

By considering simultaneously the equations of motion of the physical system and of the non-physical adjoint system, we introduce a general form of Noether's theorem by constructing a "dual Lagrangian" functional with a corresponding invariant of motion which preserves its value along the trajectories of combined physical and unphysical systems. The statement of invariance of this functional reduces to the classical statement of Noether's theorem if the system is self-adjoint; some possible generalizations are indicated. Applications to continuum mechanics are discussed within the framework of Noble's dual variational formulation.

Introduction

We introduce a k-parameter $(\varepsilon_1, \varepsilon_2, \ldots, \varepsilon_k)$ continuous group $\mathbf{G}_k$ of transformations

$$(10.1) \qquad \begin{cases} x' = f(x, y, \varepsilon), \\ y' = g(x, y, \varepsilon); \end{cases}$$

i.e., $x'^i = f^i(\mathbf{x}, \mathbf{y}, \epsilon_1, \epsilon_2, \ldots, \epsilon_k)$, $i = 1, 2, \ldots, n$, $y'^j = g^j(\mathbf{x}, \mathbf{y}, \epsilon_1, \epsilon_2, \ldots, \epsilon_k)$, $i = 1, 2, \ldots, m$, satisfying the condition

$$f^i(\mathbf{x}, \mathbf{y}, 0) = x^i,$$

$$g^i(\mathbf{x}, \mathbf{y}, 0) = y^i.$$

The group $\mathbf{G}_k$ is required to leave the Lagrangian integral invariant, that is,

$$(10.2) \qquad \int_\Omega \mathscr{L}(\mathbf{x}, \mathbf{y}, \mathbf{y}_\mathbf{x})\, d\omega = \int_{\Omega'} \mathscr{L}(\mathbf{x}', \mathbf{y}', \mathbf{y}'_{\mathbf{x}'})\, d\omega',$$

where $d\omega' = J(\mathbf{x}'/\mathbf{x})\, d\omega$, $J(\mathbf{x}'/\mathbf{x})$ denoting the Jacobian of the transformation.

The infinitesimal generators of the transformation are defined by

$$\xi_r{}^i(\mathbf{x}, \mathbf{y}) = \lim_{\epsilon \to 0} \frac{\partial f^i}{\partial \epsilon_r} \quad \eta_s{}^i(\mathbf{x}, \mathbf{y}) = \lim_{\epsilon \to 0} \frac{\partial g^j}{\partial \epsilon_s}.$$

We define the directional derivative operator

$$D_i = \frac{\partial}{\partial x_i} + y^j_{,i}\frac{\partial}{\partial y^j} + y^j_{,il}\frac{\partial}{\partial y^j_{,l}}, \qquad i = 1, 2, \ldots, n,$$

and following Eisenhart [26], the infinitesimal transformation operators are defined by the form

$$X_r = \xi_r{}^i\frac{\partial}{\partial x^i} + \eta_r{}^j\frac{\partial}{\partial y^j}, \qquad r = 1, 2, \ldots, k$$

(summation convention is tacitly assumed).
Ibrahimov [21] effects some economy in notation and arguments, by writing the infinitesimal operator in the form

$$(10.3)\quad \begin{cases} \tilde{X}_r = X_r + \xi^k_{r,i} + \mu_r\dfrac{\partial}{\partial(d\omega)}, & r = 1, 2, \ldots, k, \\ \mu_r = D_i(\xi_r{}^i)\, d\omega. \end{cases}$$

An infinitesimal variation of the Lagrangian integral is given:

$$(10.4)\qquad X_r(\mathcal{L}\, d\omega) \equiv 0.$$

Denoting Fréchet derivatives of the Lagrangian integral by $\delta I_\Omega/\delta y^j$, where, in our case,

$$(10.5)\qquad \frac{\delta I_\Omega}{\delta y^j} \equiv \frac{\partial}{\partial y^j} - D_i\frac{\partial}{\partial y^i_{,i}}$$

we can follow the formal presentation of Ibrahimov obtaining the following form of Noether's theorem:

$$(10.6)\qquad (\eta_r - y^l_{,j}\xi_r{}^j)\frac{\delta I_\Omega}{\delta y^i} + D_i(A_r{}^i) = 0, \qquad r = 1, 2, \ldots, k,$$

where

$$(10.6^a)\qquad A_r{}^i = (\eta_r{}^l - y^l_{,j}\xi_r{}^j)\frac{\partial \mathcal{L}}{\partial y^l_{,i}} + \mathcal{L}\xi_r{}^i.$$

Furthermore the following equalities determine k-invariants of motion of the systems

$$(10.7)\qquad D_i(A_r{}^i) = 0, \qquad r = 1, 2, \ldots, k,$$

provided $\delta I_\Omega/\delta y^j = 0$, $j = 1, 2, \ldots, m$.

A slightly more general form, involving multiple integrals, exists in the literature (see [21]).

Before commenting on numerous applications of this form of Noether's theorem, let us state its limitations. It is generally assumed that the equations of motion correspond to the vanishing of the Fréchet derivative of the functional I_Ω, tacitly implying the Fréchet differentability of I_Ω, the actual motion of the system corresponding to the generalized Euler-Lagrange equations. Of course a treatment using the Hamilton-Jacobi theory may be substituted. This approach to the Noether's theory can be found in the monograph of Rund[27]. A more modern version of Noether's theorem using the concepts from differential geometry and topological dynamics is given in an elegant paper of Trautman [28]. Numerous other versions of the original statements exist (see[29]).

Applications of the original version as given in [20] or of slightly modified statements abound in the literature, starting with a widely quoted (1921) paper by Bessel-Hagen[30] on applications to electrodynamics, and continuing until a recent flood of articles on applications to Korteweg-de Vries waves and solution phenomena. See, for example, [31]. Applications of Noether's theorem have been published in almost every area of classical physics, such as modern mechanics [32], optics[33], hydrodynamics[34], electromagnetic theory[30], as well as quantum mechanics, particle physics, and relativity. Since in this work, the emphasis will be on applications to classical continuum mechanics, references to other areas will be omitted.

11. A Simplified Statement of Noether's Theorem in the One-Dimensional Case

All generalizations are best presented in the simplest possible context and only then raised to the most generalized level, in the opinion of some mathematicians, including the author. Rather than attempting a complete generalization—simultaneously treating the replacement of the Lagrangian by an arbitrary functional in *n*-dimensions, completely abandoning the idea of its invariance, replacing a single functional by at least two, and simultaneously treating the motion of the physical- and the un-physical-adjoint system, moreover replacing inner products by arbitrary bilinear forms (which is the aim of this paper) and of course making the whole process completely incomprehensible, even to the initiated readers—we shall proceed cautiously, restricting ourselves to a one-dimensional case, and offer first a modest generalization in which the Lagrangian is replaced by a functional involving two basic state- and momentum-dependent variables. A convenient notation which is suitable to our purpose was introduced by Desloge and Karch [48], who offered in their expository article an elementary (variational) proof of the one-dimensional version of Noether's theorem (with a one-parameter group **G**). We consider a classical system whose motion is determined by the Lagrangian $\mathscr{L}(t, q, \dot{q})$, where t denotes the time and $\mathbf{q} = \{q_1, q_2, \ldots, q_n\}$ is the n-dimensional vector of generalized displacements of the system. We now introduce a single-parameter family of differentiable transformations

$$\begin{cases} T = T(t, \mathbf{q}, \dot{\mathbf{q}}, \epsilon), \\ Q = Q(t, \mathbf{q}, \dot{\mathbf{q}}, \epsilon), \qquad (\cdot = d/dt), \end{cases}$$

such that

$$T(t, \mathbf{q}, \dot{\mathbf{q}}, 0) = t,$$

$$Q_i(t, \mathbf{q}, \dot{\mathbf{q}}, 0) = q_i, \qquad i = 1, 2, \ldots, n.$$

We denote by

$$\Phi = \lim_{\epsilon \to 0} \int_0^t \frac{\partial}{\partial \epsilon} \{\mathscr{L}[(T(\hat{t}, \mathbf{q}, \dot{\mathbf{q}}, \epsilon), Q(\hat{t}, \mathbf{q}, \dot{\mathbf{q}}, \epsilon), \dot{Q}(\hat{t}, \mathbf{q}, \dot{\mathbf{q}}, \ddot{\mathbf{q}}, \epsilon)] \cdot \dot{T}(\hat{t}, \mathbf{q}, \dot{\mathbf{q}}, \ddot{\mathbf{q}}, \epsilon)\}\, d\hat{t},$$

$$\tau = \lim_{\epsilon \to 0} \frac{\partial T(t, \mathbf{q}, \dot{\mathbf{q}}, \epsilon)}{\partial \epsilon}, \qquad (11.1)$$

provided these limits exist. Then the theorem of Noether simplifies to the form

(11.2) $K(t, \mathbf{q}, \dot{\mathbf{q}}) \equiv$ constant along any trajectory of motion.

where

$$K = \mathscr{L}\tau + \sum_{i=1}^{n} \left(\frac{\partial \mathscr{L}}{\partial \dot{q}_i} (\psi_i - \dot{q}_i \tau) \right) - \Phi \qquad (11.2^a)$$

(see [47] for an elementary proof.)

We intend to generalize this simple form of Noether's theorem (i.e., relations (10.1) - (10.7) to systems whose equations of motion are not derivable from a single Lagrangian functional.

While many applications of Noether's theorem to continuum mechanics exist in the literature (for example, [32]), the absence of the classical form of the action integral in many problems of mechanics of continua has prevented a more widespread use of applications. In many physically realistic models of engineering systems the motion of the system is controlled by a differential equation, such that the corresponding differential operator is non-self-adjoint in the appropriate Sobolëv space topology. The introduction of a single (Lagrangian) functional which determines both the motion of the physical system and the motion of the adjoint(non-physical) system has been proposed by the author in [24], utilising Noble's theory (see [35] for a comprehensive exposition). This basic theory will be used in conjunction with Noether's arguments to derive some invariants of motion for a fairly general class of systems. Before we derive the basic one-dimensional version of our theorem, for the sake of clarity we shall offer a simple version of duality in the sense of Noble and extend Noble's ideas to incorporate generalizations introduced by Herrera, and a variant of Noble's theory suggested by the author. For original papers of Noble see [7]. For a theoretical justification of Noble's ideas see the excellent exposition of Nashed in [36]. A generalization of Noble's duality to a more abstract setting was given by Herrera and Sewell [37]. Also see Gurtin [38], Herrera & Bielak[40].

This article does not contain a full exposition of Noble's duality. Such an exposition is offered in the monograph of Arthurs [35].

12. Transformations of Variational Problems

The Euler–Lagrange equations in the classical form could represent a critical point of a map between two Hilbert spaces. Let H_q and H_p be Hilbert spaces with respective inner products $(\, , \,)$ and $\langle \, , \, \rangle$; let T be a linear map $T: H_q \to H_p$, let the domain of T be dense in H_q, let T^* be the formal adjoint of T, which is a linear map from H_p to H_q. Let $\Phi(q, p)$ be a functional of the form: $\Phi(p, q) = \langle Tq, p \rangle - W(p, q)$, $W: H_q \times H_p \to \mathbf{R}$. The functional W is assumed to be twice Fréchet differentiable. Then choosing arbitrary vectors

$\hat{q} \in H_q$, $\hat{p} \in H_p$ we can formally represent the variations of Φ by the series $\Phi(\hat{q} + \delta q, \hat{p} + \delta p) = \Phi(\hat{q}, \hat{p}) + \delta\Phi + \delta^2\Phi + o(\delta^3)$. This expression can be written in a more precise form by

$$(12.1) \qquad \delta\Phi = \left(\delta q, \frac{\delta\Phi(q, p)}{\delta q}\right)\Big|_{\substack{q=\hat{q}\\ p=\hat{p}}} + \left\langle \delta p, \frac{\delta\Phi(q, p)}{\delta p}\right\rangle\Big|_{\substack{q=\hat{q}\\ p=\hat{p}}},$$

$q = \hat{q} + \epsilon\xi$, $p = \hat{p} + \epsilon\eta$ (ϵ a "small" positive real number), $\delta q = \epsilon\xi$, $\delta p = \epsilon\eta$, and

$$\delta^2\Phi = -\frac{1}{2}\left(\delta p, \frac{\partial^2\Phi}{\partial p^2}\delta p\right) + \frac{1}{2}\left\langle \delta p, \left(T - \frac{\partial^2\Phi}{\partial p\,\partial q}\delta q\right)\right\rangle$$
$$+ \frac{1}{2}\left\langle \delta q, \left(T^* - \frac{\partial^2\Phi}{\partial q\,\partial p}\right)\delta p\right\rangle - \frac{1}{2}\left\langle \delta q, \frac{\partial^2\Phi}{\partial q^2}\delta q\right\rangle - \frac{K}{2}[\delta q, \delta p],$$

where the last term satisfies the "relative smallness" condition resulting from Fréchet differentiability of Φ.

The formalism of Noble introduces the Hamiltonian $W(q, p)$, whose form is to be determined and a Lagrangian

$$(12.2) \qquad \mathscr{L} = \langle p, Tq\rangle - W = (T^*p, q) - W,$$

such that the equations of the system can be written in the form

$$(12.3) \qquad \frac{\delta\mathscr{L}}{\delta h} = \begin{pmatrix} Aq - \delta W/\delta p \\ A^*p - \delta W/\delta q \end{pmatrix} = 0,$$

where $h = \binom{p}{q}$ is considered as a vector in the space $H_q \oplus H_p$. The author suggested in [32] that problems which are not derivable from variational principles in the sense of Tonti [49] can still be represented by Noble's formalism. However, spaces containing the dual or adjoint displacement and momentum variables have to be introduced to derive a generalization of Eqs. (12.3). Let us illustrate the general idea by offering a simple example. Again, to avoid unnecessary complexity, we consider a familiar one-dimensional case. Consider the second order differential system

$$(12.3\text{A}) \qquad \begin{cases} m\ddot{u} + \mu(t)\,\dot{u} + k(t)\,u = f(t), \\ u(0) = \dot{u}(0) = 0, \end{cases}$$

on the time interval $0 \leq t \leq 1$. Introducing a function $v \in C^1[0, 1]$ satisfying $v(1) = \dot{v}(1) = 0$, we interpret Eq. (A) as the sufficient condition for vanishing of the Fréchet derivative of the functional $\mathscr{L}_{1,2}$, namely,

(12.4a) $$\frac{\delta \mathscr{L}_{1,2}}{\delta v} = 0,$$

where

(12.5) $$\mathscr{L}_{1,2}(u, v) = \int_0^1 [(m\ddot{u} + \mu(t)\,\dot{u} + K(t)\,u - f(t))\,v(t)]\,dt,$$

$$\mathscr{L}_{1,2} = \langle (Lu - f), v \rangle.$$

The adjoint equation

$$m\ddot{v} - \frac{d}{dt}(\mu v) + K(t)\,v = 0,$$

with $v(1) = \dot{v}(1) = 0$, corresponds to the condition

(12.4b) $$\frac{\delta \mathscr{L}_{1,2}}{\delta u} = 0.$$

Conditions (12.4a), (12.4b) can be written in the Euler-Lagrange form

(12.4c) $$\frac{\partial \mathscr{L}_{1,2}}{\partial v} - \frac{d}{dt}\left(\frac{\partial \mathscr{L}_{1,2}}{\partial \dot{v}}\right) = 0,$$

and

(12.4d) $$\frac{\partial \mathscr{L}_{1,2}}{\partial u} - \frac{d}{dt}\left(\frac{\partial \mathscr{L}_{1,2}}{\partial \dot{u}}\right) = 0.$$

The functional (12.5) can be written in a "symmetric" form

$$\mathscr{L}_{1,2}(t, u, v, \dot{u}, \dot{v}) = \int_0^1 \left[-m\dot{u}\dot{v} + \frac{1}{2}(\dot{u}v\mu) - u\frac{d}{dt}(\mu v) + kuv - f(t)\,v\right] dt.$$

Note. In fact the initial conditions (12.3a) and the final conditions may be written as a derivative of a boundary integral with a discrete measure assigned to the points 0 and 1, but for the time being, all such complications are deliberately avoided.

We introduce the generalized momenta

$$p_1 = \frac{\partial \mathscr{L}_{1,2}(t, u, \dot{u}, v, \dot{v})}{\partial \dot{u}},$$

$$p_2 = \frac{\partial \mathscr{L}_{1,2}(t, u, \dot{u}, v, \dot{v})}{\partial \dot{v}}.$$

Conditions (c), (d) can be replaced by an equivalent system of Hamilton's "canonical" equations:

$$
(12.6)\qquad \begin{cases} \dfrac{\partial \mathscr{H}}{\partial u} - \dfrac{d}{dt}(p_2) = 0, \\ \dfrac{\partial \mathscr{H}}{\partial v} - \dfrac{d}{dt}(p_1) = 0, \\ \dfrac{\partial \mathscr{H}}{\partial p_1} + \dfrac{d}{dt}(v) = 0, \\ \dfrac{\partial \mathscr{H}}{\partial p_2} + \dfrac{d}{dt}(u) = 0, \end{cases}
$$

where the Hamiltonian $\mathscr{H}$ is given by

$$
(12.7)\qquad \mathscr{H} = p_1\dot{v} + p_2\dot{u} - \mathscr{L}_{1,2}(t, u, \dot{u}, v, \dot{v}).
$$

Following the original remarks of Noble[7], we comment that in a suitable Hilbert space setting the operators d/dt, $-d/dt$ may be replaced by linear operators A, A^*, $A: H_1 \to H_2$, $A^*: H_2 \to H_1$ while preserving formally all properties of the Lagrangian and Hamiltonian formalism.

This idea will be applied directly to Noether's theory.

In the next section we shall introduce only a moderate generalization of Noether's theorem, using only a simple Lagrangian, and utilizing the property of only a single-parameter group of transformations.

13. A Simple Extension of Noether's Theorem

We introduce a bilinear form

$$
(13.1)\qquad \mathscr{L}_{1.2} = (t, \mathbf{u}, \mathbf{v}, \dot{\mathbf{u}}, \dot{\mathbf{v}})
$$

such that the behavior of the physical system which we model mathematically is represented by the relation

$$
(13.2)\qquad \frac{\delta \mathscr{L}_{1.2}}{\delta \mathbf{v}} = 0, \qquad \text{where} \qquad \frac{\delta}{\delta \mathbf{v}} \equiv \begin{pmatrix} \dfrac{\delta}{\partial v_1} \\ \vdots \\ \dfrac{\partial}{\partial v_n} \end{pmatrix}.
$$

As before, $\mathbf{u}(t)$ is the state vector variable of the physical system, $\mathbf{v}(t)$ is the adjoint (un-physical) vector, and t represents time.

A family of one-parameter smooth transformations is defined by equations of type :

$$U = U(t, u, v, \dot{u}, \dot{v}, \epsilon),$$
$$V = V(t, u, v, \dot{u}, \dot{v}, \epsilon),$$
$$T = T(t, u, v, \dot{u}, \dot{v}, \epsilon), \qquad (13.1)$$

We require that

$$u(t) = U(t, u, v, \dot{u}, \dot{v}, 0),$$
$$v(t) = V(t, u, v, \dot{u}, \dot{v}, 0),$$
$$t = T(t, u, v, \dot{u}, \dot{v}, 0),$$

i.e., when $\epsilon = 0$ the coordinate systems u, v, and U, V coincide, respectively, and the time scales t and T are identical.

We introduce the following additional quantities:

$$\left.\begin{aligned} \tau &= \lim_{\epsilon\to 0} \frac{\partial T(t, q, \dot{q}, \epsilon)}{\partial \epsilon}, \\ \eta_i &= \lim_{\epsilon\to 0} \frac{\partial U_i(t, u, v, \dot{u}, \dot{v}, \epsilon)}{\partial \epsilon}, \\ \psi_i &= \lim_{\epsilon\to 0} \frac{\partial V_i(t, u, v, \dot{u}, \dot{v}, \epsilon)}{\partial \epsilon}. \end{aligned}\right\} \qquad (13.2)$$

We shall denote by $\mathscr{L}$ the "mutual Lagrangian" $\mathscr{L}_{1,2}(t, u, v, \dot{u}, \dot{v})$ obeying the equations

$$(13.3^a) \qquad \frac{\partial \mathscr{L}}{\partial v} - \frac{d}{dt}\left(\frac{\partial \mathscr{L}}{\partial \dot{v}}\right) = 0,$$

$$(13.3^b) \qquad \frac{\partial \mathscr{L}}{\partial u} - \frac{d}{dt}\left(\frac{\partial \mathscr{L}}{\partial \dot{u}}\right) = 0,$$

which are the equations of motion of the physical system and of the adjoint system, respectively. We shall use the following notation: $\hat{\mathscr{L}}$ will denote $\mathscr{L}(T, U, V, \dot{U}, \dot{V}, \epsilon)$, $\circ$ will denote differentiation with respect to T, and $\cdot$ with respect to t. P_1 and P_2 denote $\partial\hat{\mathscr{L}}/\partial\mathring{U}$, $\partial\hat{\mathscr{L}}/\partial\mathring{V}$, respectively. $\hat{H}$ will denote the Hamiltonian defined by relation (12.7) with respect to $\hat{\mathscr{L}}$. The following relations are readily established.

$$\left.\begin{aligned} \left.\frac{\partial \dot{T}}{\partial \epsilon}\right|_{\epsilon=0} &= \tau, \\ \dot{T}\,|_{\epsilon=0} &= 1, \\ \dot{U}\,|_{\epsilon=0} &= \dot{u}\,|_{\epsilon=0}, \end{aligned}\right\} \qquad (13.4)$$

$$\left.\begin{aligned} &\dot{V}\,|_{\epsilon=0} = \dot{v}\,|_{\epsilon=0}\,, \\ &\frac{\partial \dot{u}}{\partial \epsilon}\Big|_{\epsilon=0} = \dot{\eta}, \\ &\frac{\partial \dot{V}}{\partial \epsilon}\Big|_{\epsilon} = \psi, \end{aligned}\right\} \tag{13.5}$$

$$\left.\begin{aligned} &\mathcal{L}\,|_{\epsilon=0} = \hat{\mathcal{L}}\,|_{\epsilon=0}\,, \\ &H\,|_{\epsilon=0} = \hat{H}\,|_{\epsilon=0}\,, \end{aligned}\right\} \tag{13.6}$$

$$\left.\begin{aligned} &\frac{\partial \mathring{U}_i}{\partial \epsilon}\Big|_{\epsilon=0} = \frac{\partial}{\partial \epsilon}\left(\frac{\dot{U}_i}{\mathring{T}}\right)\Big|_{\epsilon=0} = \dot{\eta}_i - \dot{u}_i\dot{\tau}, \\ &\frac{\partial \mathring{V}_i}{\partial \epsilon}\Big|_{\epsilon=0} = \psi_i - \dot{v}_i\dot{\tau}, \end{aligned}\right\} \tag{13.7}$$

$$\begin{aligned} &\frac{\partial \mathcal{L}}{\partial t}\Big|_{\epsilon=0} \quad \frac{\partial \hat{\mathcal{L}}}{\partial T}\Big|_{\epsilon=0}, \\ &\frac{\partial \hat{\mathcal{L}}(T, U, V, \ldots)}{\partial U_i}\Big|_{\epsilon=0} = \frac{\partial \mathcal{L}}{\partial u_i}\Big|_{\epsilon=0}, \end{aligned} \tag{13.8}$$

$$\left.\begin{aligned} &P_{1_i}\,|_{\epsilon=0} = \frac{\partial \hat{\mathcal{L}}}{\partial \mathring{U}_i}\Big|_{\epsilon=0} = p_{1_i}\,|_{\epsilon=0}\,\frac{\partial \mathcal{L}}{\partial \dot{u}_i}\Big|_{\epsilon=0}, \\ &\frac{\partial \hat{\mathcal{L}}}{\partial V_i}\Big|_{\epsilon=0} = \frac{\partial \mathcal{L}}{\partial v_i}\Big|_{\epsilon=0}, \\ &P_{2_i}\,|_{\epsilon=0} = \frac{\partial \hat{\mathcal{L}}}{\partial \mathring{V}_i}\Big|_{\epsilon=0} = p_{2_i}\,|_{\epsilon=0} = \frac{\partial \mathcal{L}}{\partial \dot{v}_i}\Big|_{\epsilon=0}. \end{aligned}\right\} \tag{13.9}$$

We define the functional Φ, such that

$$\frac{\partial}{\partial \epsilon}\{\hat{\mathcal{L}}(U, \mathring{U}, V, \mathring{V}, T)\cdot \mathring{T}\}\,|_{\epsilon=0} = \Phi \tag{13.10}$$

and a functional

$$\Lambda = \mathcal{L}\tau + \sum_{i=1}^{n} p_{1_i}(\eta_i - \dot{u}_i\tau) + \sum_{i=1}^{n} p_{2_i}(\psi_i - \dot{v}_i\tau), \tag{3.11}$$

in analogy with formulas (8.8) and (8.6) of Section 1 of this chapter, and proceed to show that $\Lambda - \Phi$ is an invariant of the motion of a composite system described by Eqs.(13.3[a]) and

(13.3[b]) that is of the original equation and its dual

Denoting by $\langle \boldsymbol{\alpha}, \boldsymbol{\beta} \rangle = \sum_{i=1}^{n} \alpha_i \beta_i$, we compute the following expressions:

$$\Phi = \frac{\partial}{\partial \epsilon} \{\hat{\mathscr{L}}(U, \mathring{U}, V, \mathring{V}, T) \cdot \mathring{T}\}_{\epsilon=0}$$

$$= \left[\frac{\partial \hat{\mathscr{L}}}{\partial T} \frac{\partial T}{\partial \epsilon} + \frac{\partial \hat{\mathscr{L}}}{\partial U} \frac{\partial U}{\partial \epsilon} + \frac{\partial \hat{\mathscr{L}}}{\partial V} \frac{\partial V}{\partial \epsilon} + \frac{\partial \hat{\mathscr{L}}}{\partial \mathring{U}} \frac{\partial \mathring{U}}{\partial \epsilon} + \frac{\partial \hat{\mathscr{L}}}{\partial \mathring{V}} \frac{\partial \mathring{V}}{\partial \epsilon} + \hat{\mathscr{L}} \frac{\partial \mathring{T}}{\partial \epsilon}\right]_{\epsilon=0}$$

$$= \left[\left\langle \frac{\partial \mathscr{L}}{\partial t}, \tau \right\rangle + \left\langle \frac{\partial \mathscr{L}}{\partial u}, \eta \right\rangle + \left\langle \frac{\partial \mathscr{L}}{\partial v}, \psi \right\rangle + \langle p_1, \dot{\eta} \rangle + \langle p_2, \dot{\psi} \rangle + \mathscr{L}\dot{\tau}\right]_{\epsilon=0}$$

$$- [\langle p_1 \dot{u}, \tau \rangle + \langle p_2, \dot{u}\dot{\tau} \rangle]_{\epsilon=0}, \qquad (13.12)$$

$$\frac{d}{dt} (\Lambda)_{\epsilon=0}$$

$$= \frac{d}{dt} \{\mathscr{L}\tau + [\langle p_1, (\eta - \dot{u}\tau) \rangle + \langle p_2, (\psi - \dot{v}\tau) \rangle]\}_{\epsilon=0}$$

$$= \left\{\frac{\partial \mathscr{L}}{\partial t} \tau + \frac{\partial \mathscr{L}}{\partial u} \tau + \frac{\partial \mathscr{L}}{\partial v} \dot{v}\tau + p_1 \ddot{u}\tau + p_2 \ddot{v}\tau + \mathscr{L}\dot{\tau}\right.$$

$$+ \langle \dot{p}_1, \eta \rangle + \langle p_1, \dot{\eta} \rangle - \langle \dot{p}_1, \dot{u}\tau \rangle - \langle p_1, \ddot{u}\tau \rangle - \langle p_1, \dot{u}\dot{\tau} \rangle + \langle \dot{p}_2, \psi \rangle$$

$$\left. + \langle p_2 \dot{\psi} \rangle - \langle \dot{p}_2, \dot{v}\tau \rangle - \langle p_2, \ddot{v}\tau \rangle - \langle p_2, \dot{v}\dot{\tau} \rangle \right\}_{\epsilon=0}. \qquad (13.13)$$

Concelling out terms by observing that

$$\frac{\partial \mathscr{L}}{\partial u} \dot{u}\tau = \dot{p}_1 \dot{u}\tau$$

$$\frac{\partial \mathscr{L}}{\partial v} \dot{v}\tau = \dot{p}_2 \dot{v}\tau$$

we obtain after rearranging of terms

$$\frac{d}{dt} (\Lambda - \Phi) = 0. \qquad (13.14)$$

Hence $(\Lambda - \Phi)$ is constant along the joint equations of motion of both systems (13.13[a]), and (13.13[b]).

This is formalized as a theorem.

THEOREM *Let $\mathscr{L}(t, u, v, \dot{u}, \dot{v})$ satisfy the equations*

$$\frac{\partial \mathscr{L}}{\partial u} - \frac{d}{dt}\left(\frac{\partial \mathscr{L}}{\partial \dot{u}}\right) = 0, \qquad (13.15)$$

$$\frac{\partial \mathscr{L}}{\partial v} - \frac{d}{dt}\left(\frac{\partial \mathscr{L}}{\partial \dot{v}}\right) = 0. \qquad (13.16)$$

Let a family of U, V, T be defined by a one-parameter family of differentiable transformations (13.1) satisfying the conditions (13.1). τ, ψ, η are defined as the limits (13.2).

Then the functional $G(U,u,V,v,T,t,\dot{U},\dot{u},\dot{V},\dot{v},\dot{T}) = \Phi - \Lambda$, where Φ, Λ are defined by Eqs. (13.10), (13.11)

If the system is self-adjoint with respect to the product < , >, then $u(t) = v(t)$, Eqs. (13.15), (13.16) are identical, and the theorem reduces to the usual statement of Noether's theorem for the system discussed in Section 1 of this chapter.

We note that in that case the "mutual Lagrangian" $\mathscr{L}_{1,2}(t, u, (v = u), \dot{u}, (\dot{v} = \dot{u}))$ is twice the "usual" Lagrangian $\mathscr{L}(t, u, \dot{u})$. We perform an easy check on the meaning of this version of the theorem by considering the simplest case.

Let $T = \tau + \epsilon$, $\mu = 0$, and $u = v$, $\tau \equiv 1$. Then $\dot{\Phi} = 0$ and $\Phi \equiv$ const, $\eta_i \equiv 0$, $\psi_i \equiv 0$, $\dot{\tau} \equiv 0$, and we compute $L - \sum_{i=1}^{3} p_i \dot{u}_i +$ constant $\equiv$ constant. Hence $L - \sum_{i=1}^{3} p_i \dot{u}_i = H =$ constant. Here L denotes the "usual" Lagrangian, and H the Hamiltonian of the motion.

Example of Application. We consider a system defined by the "mutual" Lagrangian density function

$$\mathscr{L}(\tau, x, y, \dot{x}, \dot{y}) = \sum_{i=1}^{3} [-\kappa_i x_i y_i + m(\tau)\, \dot{x}_i \dot{y}_i + \mu_i \dot{x}_i y_i - f_i(t)\, y_i],$$

where κ_i, μ_i are constants. We introduce a new coordinate system

$$X_i = x_i + \epsilon \alpha_i t,$$
$$Y_i = y_i + \epsilon \beta_i t,$$
$$T = t + t\phi(\epsilon),$$

where α_i, β_i are constants satisfying the relation $\alpha^2 + \beta^2 = 1$. $\phi \in C^1[0.1]$, $\phi(0) = 0$, $\phi'(0) = 1$. We compute the quantities Λ and Φ, which satisfy the relations (13.10 and (13.11). In particular

$$\Lambda = \tfrac{1}{2}\left\{\mathscr{L}\tau + \sum_{i=1}^{3} (p_{1_i}(\eta_i - \dot{x}\tau) + p_{2_i}(\psi_i - \dot{y}_i\tau))\right\}.$$

We substitute

$$\eta_i = \alpha_i t, \qquad \psi_i = \beta_i t,$$
$$\tau = t\phi'(0) = t, \qquad \dot\tau = 1,$$
$$p_{1_i} = m\dot y_i - \mu_i \dot y_i, \qquad p_{2_i} = m\dot x_i + \mu_i \dot x_i,$$

obtaining

$$\Lambda = \mathscr{L}t + \sum_{i=1}^{3} [\dot y_i(m - \mu_i)(\alpha_i t - \dot x_i) + \dot x_i(m + \mu_i)(\beta_i t - \dot y_i)].$$

$$\Phi = \int_0^t \left[t\frac{\partial \mathscr{L}}{\partial t} + \sum_{i=1}^{3}\left(\alpha_i t\frac{\partial \mathscr{L}}{\partial x_i} + \beta_i t\frac{\partial \mathscr{L}}{\partial y_i} + p_{1_i}\alpha_i + p_{2_i}\beta_i\right) + \mathscr{L}\dot\tau\right] dt$$

$$= \int_0^t \left[t\left(\sum_{i=1}^{3}(m x_i y_i - \dot f_i y_i)\right) + \sum_{i=1}^{3} (\kappa_i\alpha_i y_i t - \kappa_i\beta_i x_i t - \alpha_i(m - \mu_i)\dot y_i \right.$$
$$\left. + \beta_i(m + \mu_i)\dot x_i) + \mathscr{L}\right] dt + \text{constant}.$$

$(\Lambda - \Phi)$ is a constant of motion along an arbitrary pair of trajectories of the physical and un-physical motions described by the equations:

$$\frac{d}{dt}(m\dot x_i) + \mu_i \dot x_i + \kappa_i x_i = f,$$
$$\frac{d}{dt}(m\dot y_i) - \mu_i \dot y_i + \kappa_i y_i = 0,$$

with zero initial conditions assigned to $x_i(t)$, and zero final conditions assigned to $y_i(t)$. ∎

14. A Group of Transformations Which Leaves "the Mutual Lagrangian" $\mathscr{L}_{1,2}$ Invariant

We write out in full the condition for invariance of the Lagrangian $\mathscr{L}_{1,2} = \mathscr{L}$. The full equation $\dot\Phi - \dot\Lambda = 0$ is represented by the limit:

$$\frac{\partial \mathscr{L}}{\partial t}\tau + \left\langle \frac{\partial \mathscr{L}}{\partial u}, \eta\right\rangle + \left\langle \frac{\partial \mathscr{L}}{\partial v}, \psi\right\rangle + \langle p_1, \dot\eta\rangle + \langle p_2, \dot\psi\rangle - \langle p_1 \dot u, \dot\tau\rangle$$
$$- \langle p_2 \dot v, \dot\tau\rangle - \frac{d}{dt}(\mathscr{L}\tau) + \frac{d}{dt}\{\langle p_1, (\eta - \dot u\tau)\rangle + \langle p_2, (\psi - \dot v\tau)\rangle\} = 0. \tag{14.1}$$

We could regard $\dot u$, $\dot v$ as arbitrary functions, and represent p_1, p_2 in terms of $u, \dot u, v, \dot v$. Eq. 14.1 is then regarded as a partial differential equation of first order in τ, ψ and η, which must be satisfied for arbitrary choices of $\dot u$, $\dot v$. The usual technique consists of equating to zero different powers of u. Since we deal with variables u and v simultaneously, the temptation to regard both $\dot{u}$ and $\dot{v}$ as independent variables, copying the usual procedure, must be resisted. $\dot{u}$ can be regarded as independent of u, and $\dot{v}$ as independent of v, but u and v must be adjoints of each other in the topology assigned to the problem, and $\dot{u}$ and $\dot{v}$ cannot be regarded as independent of each other.

EXAMPLE. Consider the Emden equation with a dissipative term given by the Lagrangian

$$\mathscr{L} = \mathscr{L}_{1,2}(t, u, v) = t^2\left(\frac{\dot{u}\dot{v}}{2} - \frac{(uv)^3}{6}\right) + \frac{t^2\dot{u}v}{2}.$$

The one-parameter family of transformations is given by

$$\begin{aligned} T &= t + \epsilon\tau(t, u, v), \\ U &= u + \epsilon\eta(t, uv), \\ V &= v + \epsilon\psi(t, u, v). \end{aligned}$$

We have:

$$\begin{aligned} \frac{\partial\mathscr{L}}{\partial t} &= t[\dot{u}\dot{v} - (uv)^3 + \dot{u}v], \\ \frac{\partial\mathscr{L}}{\partial u} &= -\frac{1}{2}(3t^2(uv)^2 v + t^2\dot{v}), \\ \frac{\partial\mathscr{L}}{\partial v} &= \frac{1}{2}(-3t^2(uv)^2 u + t^2\dot{u}), \\ p_1 &= \frac{\partial\mathscr{L}}{\partial\dot{u}} = \frac{1}{2}(t^2\dot{v} + t^2 v), \\ p_2 &= \frac{\partial\mathscr{L}}{\partial\dot{v}} = \frac{1}{2}(t^2\dot{u} - t^2 u). \end{aligned}$$

The invariance relation for $\mathscr{L}$ is given by

$$\frac{\partial\mathscr{L}}{\partial t}\tau + \frac{\partial\mathscr{L}}{\partial u}\eta + \frac{\partial\mathscr{L}}{\partial\dot{u}}(\dot{\eta} - \dot{u}\dot{\tau}) + \frac{\partial\mathscr{L}}{\partial v}\psi + \frac{\partial\mathscr{L}}{\partial v}(\dot{\psi} - \dot{v}\dot{\tau}) + \mathscr{L}\dot{\tau} = 0,$$

which becomes:

$$\begin{aligned} &t[\dot{u}\dot{v} - (uv)^3 + \dot{u}v]\,\tau - \tfrac{1}{2}(3t^2(uv)^2 v + t^2\dot{v})\,\eta + \tfrac{1}{2}(t^2\dot{v} + t^2 v)\cdot(\dot{\eta} - \dot{u}\dot{\tau}) \\ &\quad + \tfrac{1}{2}(-3t^2(uv)^2 u + t^2\dot{u}) + \tfrac{1}{2}(t^2\dot{u} - t^2 u)\,(\dot{\psi} - \dot{v}\dot{\tau}) \\ &\quad + \dot{\tau}t^2\left[\left(\frac{\dot{u}\dot{v}}{2} - \frac{uv^3}{6}\right) + \frac{\dot{u}v}{2}\right] = 0. \end{aligned}$$

Replacing

$$\dot{\tau} \quad \text{by} \quad \frac{\partial \tau}{\partial t} + \frac{\partial \tau}{\partial u}\dot{u} + \frac{\partial \tau}{\partial v}\dot{v},$$

$$\dot{\eta} \quad \text{by} \quad \frac{\partial \eta}{\partial t} + \frac{\partial \eta}{\partial u}\dot{u} + \frac{\partial \eta}{\partial v}\dot{v},$$

$$\dot{\psi} \quad \text{by} \quad \frac{\partial \psi}{\partial t} + \frac{\partial \psi}{\partial u}\dot{u} + \frac{\partial \psi}{\partial v}\dot{v},$$

we obtain the following equation:

$$\begin{aligned}
&\left[-t(uv)^3\tau - \frac{3}{2}t^2u^2v^3 + \frac{1}{2}t^2v\frac{\partial \eta}{\partial t} - \frac{2}{2}t^2u^3v^2 - \frac{1}{2}t^2u\frac{\partial \psi}{\partial t} - t^2\frac{uv^3}{6}\frac{\partial \tau}{\partial t}\right] \\
&\quad + \left[t\dot{u}v\tau - \frac{1}{2}t^2\dot{v}\eta + \frac{1}{2}t^2\dot{v}\left(\frac{\partial \eta}{\partial t} + \frac{\partial \eta}{\partial u}\dot{u} + \frac{\partial \eta}{\partial v}\dot{v}\right)\right. \\
&\quad - \frac{1}{2}\left(t^2v\dot{u}\frac{\partial \tau}{\partial t} + t^2v\dot{u}^2\frac{\partial \tau}{\partial u} + t^2v\dot{v}\dot{u}\frac{\partial \tau}{\partial v}\right) \\
&\quad + \frac{1}{2}t^2\dot{u}\left(1 + \frac{\partial \psi}{\partial t} + \frac{\partial \psi}{\partial u}\dot{u} + \frac{\partial \psi}{\partial v}\dot{v}\right) - \frac{1}{2}t^2\dot{u}\dot{v}\left(\frac{\partial \tau}{\partial t} + \frac{\partial \tau}{\partial u}\dot{u} + \frac{\partial \tau}{\partial v}\dot{v}\right) \\
&\quad \left. + \frac{1}{2}t^2u\dot{v}\left(\frac{\partial \tau}{\partial t} + \frac{\partial \tau}{\partial u}\dot{u} + \frac{\partial \tau}{\partial v}\dot{v}\right)\right] = 0,
\end{aligned}$$

after cancellation of some terms. We can regard the collections $\{u, v\}$, $\{\dot{u}, \dot{v}\}$ $\{\dot{u}^2, \dot{u}\dot{v}, \dot{v}^2\}$, etc., as independent of each other in the following sense: u and v depend on each other, but u can be chosen independently of $\dot{u}$, $\dot{v}$, $\dot{u}^2$, $\dot{v}^2$, etc. Hence, equating separately to zero the terms which contain u, v only, then the terms which contain $\dot{u}$, $\dot{v}$, but not $\dot{u}^2$, $\dot{v}^2$, $\dot{u}\dot{v}$, and so on, we derive a set of partial differential equations in τ, η, ψ. The highest powers of $\dot{u}$, $\dot{v}$ in the second bracket occur in the term

$$\left(-\frac{1}{2}t^2\dot{u}^2v + \frac{1}{2}t^2u\dot{v}^2\right)\frac{\partial \tau}{\partial u} + \left(-\frac{1}{2}t^2\dot{u}v^2 + \frac{1}{2}t^2\dot{u}^2\dot{v}\right)\frac{\partial \tau}{\partial v}.$$

Equating such terms to zero provides one of the equations of our system. Other equations are derived in a similar manner. There is no point in reproducing them here, except possibly to compare them with the known invariants of the Emden equation. There is no reason why $\partial\tau/\partial u$ and $\partial\tau/\partial v$ should vanish separately along the trajectory of the system. Hence the well-known first integral of the Emden equation is no longer valid for this case.

§5. An Example from Continuum Mechanics

Let us assume Hooke's law and the simplest form of dissipation of the dashpot type (the Voigt–Maxwell model), so that

$$\sigma_{ij,j} = \rho\ddot{u} + \mu\dot{u} \quad (15.1)$$

are the equations of motion, which represent Newton's second law. Hooke's law is given by the stress-strain relation

$$\sigma_{ij} = C_{ijkl}\epsilon_{kl}, \quad \text{where} \quad \epsilon_{ij} = \tfrac{1}{2}(u_{i,j} + u_{j,i}). \quad (15.1^a)$$

In the absence of the dissipative term, the Lagrangian would be a simple difference of the kinetic and potential energy

$$(15.2) \qquad \mathscr{L} = \tfrac{1}{2}(\langle\rho\dot{u}, \dot{u}\rangle - \langle C_{ijk}\epsilon_{ijk}\epsilon_k\rangle).$$

Introducing a vector field which is invariant under infinitesimal invariance of $\mathscr{L}$ leads directly to the class of conservation laws derived by different techniques by Knowles and Sternberg[29]. The presence of a dissipative term does not permit the use of the classical version of Noether's theorem.

Introducing the adjoint variable **v**, we consider the Lagrangian

$$\mathscr{L}_{1,2}(u, v) = \tfrac{1}{2}(\langle\rho\dot{u}, \dot{v}\rangle - \tfrac{1}{2}C_{ijkl}[(u_{i,j} + u_{j,i})(v_{k,l} + v_{l,k})] + \tfrac{1}{4}\mu[\langle\dot{u}, v\rangle - \langle u, \dot{v}\rangle],$$
$$(15.2^a)$$

which will be called the dual elastic Lagrangian. Let x denote the usual Cartesian coordinates. We define a family

$$(15.3) \qquad \{t, x, u(x, t), v(x, t)\} \to \{\tau, \xi, u^*(\xi, \tau), v^*(\xi, \tau), \epsilon\}$$

of class $C^2(R^4)$, satisfying Noether's condition at $\epsilon = \varnothing$ i.e., $t(\epsilon = 0) = T(\epsilon = 0)$, $x(\epsilon = 0) = \xi(\epsilon = 0)$, etc., with the invariance condition

$$(15.4) \qquad \frac{d}{d\epsilon}\left\{\int_{R^4} \mathscr{L}_{1,2}(T, \xi, u^*, v^*, \epsilon)\right\}\Bigg|_{\epsilon=0} = 0.$$

Theorem. *Let u, v be vector fields, such that*

$$\mathscr{L}_{1,2}\left(u, v, u_x, v_x, \frac{\partial u}{\partial t}, \frac{\partial v}{\partial t}, x, t\right)$$

satisfies the Euler–Lagrange condition

(15.5[a]) $$\frac{\partial \mathscr{L}_{1.2}}{\partial u} - \frac{\partial}{\partial x}\frac{\partial \mathscr{L}_{1.2}}{\partial u_x} - \frac{\partial}{\partial t}\frac{\partial \mathscr{L}_{1.2}}{\partial u_t} = 0,$$

(15.5[b]) $$\frac{\partial \mathscr{L}_{1.2}}{\partial v} - \frac{\partial}{\partial x}\frac{\partial \mathscr{L}_{1.2}}{\partial v_x} - \frac{\partial}{\partial t}\frac{\partial \mathscr{L}_{1.2}}{\partial v_t} = 0.$$

Then the dual elastic Lagrangian $\mathscr{L}_{1.2}$ is invariant under a class of C^2 transformation (15.3) if and only if it satisfies the generalized Noether conditions (15.6), (15.7) given below.

Note. The derivatives in formulas (15.5[a]), (15.5[b]) are taken in the Fréchet sense.

The invariance of the dual Lagrangian $\mathscr{L}_{1,2}$ functional is given in terms of the following equations for the Lagrangian density function L

$$\sum \frac{\partial}{\partial x_j}\left\{\frac{\partial L}{\partial u_{i,j}}\cdot(\psi_i - u_{i,j}\phi_j) + L\phi_j\right\} + \frac{\partial}{\partial x_j}\left(\frac{\partial L}{\partial v_{i,j}}(\eta_i - v_{i,j}\phi_j) + L\phi_j\right)$$
$$- \frac{\partial}{\partial t}\left\{\frac{\partial L}{\partial \dot{u}_j}\dot{u}_{jT} + \frac{\partial L}{\partial \dot{v}_j}\dot{v}_{jT}\right\} + L_T = 0, \qquad (15.6)$$

where

$$\phi_j = \frac{\partial \xi_j}{\partial \epsilon}\bigg|_{\epsilon=0},$$
$$\psi_j = \frac{\partial}{\partial \epsilon}u_j^*\bigg|_{\epsilon=0}, \qquad (15.7)$$
$$\mu_j = \frac{\partial}{\partial \epsilon}v_j^*\bigg|_{\epsilon=0},$$

The proof is identical to the one given earlier except for technical details. Obviously, integrating the expression over any region

$$\{\Omega \times [0, T]\}, \qquad \Omega \subset R^3, \quad \text{or over } \partial\Omega \times [0, T]$$

still preserves the infinitesimal invariance of the transformation (15.3). Eliminating time dependence of the functions u, v, setting $\mu \equiv 0$, hence writing $u \equiv v$, $u^* \equiv v^*$, we obtain the exact results of Knowles and Sternberg [29]

16. OTHER POSSIBLE APPROACHES

Noether's theorem basically implies that for each group of infinitesimal transformations which leaves the action integral or the generalized Lagrangian integral invariant there exists a conserved functional, that is, a functional which is invariant along any trajectory of motion (see [33]). The approach of this work replaces the classical action integral by a "dual" Lagrangian integral, basically preserving Noether's approach. The question arises what other quantities can be substituted for the action integral to produce either variational principles or conserved quantitites in the manner of Noether's theorem. In a recent article [41] Vujanovič points out that taking d'Alambert's principle in the form of the principle of virtual work leads in the manner of Noether's theorem to the formation of conserved quantities of a dynamical system. In fact, taking the virtual work formula in the form

$$\frac{d}{dt}\left(\frac{\partial L}{\partial \dot{q}_i}\delta q_i\right) - \frac{\partial L}{\partial \dot{q}_i}\delta \dot{q}_i - \frac{\partial L}{\partial q_i}\delta q_i - Q_i\delta q_i = 0, \qquad (16.1)$$

Vujanović establishes the "gauge" formula

$$\dot{\Lambda} - \dot{\Phi} = 0, \qquad (16.2)$$

where

$$\Lambda = \left(\frac{\partial L}{\partial \dot{q}}(\Delta q_i - \dot{q}_i\Delta t) + L\Delta t - \epsilon P\right), \qquad (16.3)$$

and

$$\dot{\Phi} = \Delta L + L(\Delta t)^{\cdot} + Q_i(\Delta q_i - \dot{q}_i\Delta t) - \epsilon \dot{P}, \qquad (16.4)$$

where P is the gauge function. The novelty consists in permitting the generalized force Q_i, which may not be derivable from a potential, to enter the Noether's formulation.

The corresponding partial differential equations which define conservation laws turn out to be the Killing's equations. Observing that the principle of virtual work really represents the total work of a dynamical system, it appears that one can safely copy the arguments of Vujanovič by considering abstract energy forms which generalize the principle of virtual work in various problems of continuum mechanics. In this respect the energy principles using the internal variables approach developed by Valanis and Komkov in [37] for thermodynamics of continua seem to offer a starting point. While the basic approach to the finding of invariants of motion remains the same, the introduction of internal variables allows us to consider some averages for complex materials in formulating constitutive properties. Roughly speaking the internal variables represent averages of molecular displacements.

It is clear that a microsystem can not be regarded as arbitrarily small, or infinitesimal, if the concepts of average physical quantities such as energy density, stress, entropy density, are to be physically significant in a physical medium which is made of particles. This is so, since at an atomic level these quantities are either undefined or at best badly discontinuous. In the case of polymeric materials, regarding them as dissipative thermodynamics systems, Valanis identified the internal variables as actual displacements along a typical very large molecule. Hence, in the study of polymers q_α could be identified with average displacements and corresponding generalized forces could be identified with the energy gradient $\nabla\psi = \{\partial\psi/\partial q_\alpha\}$ where ψ denotes the Helmholtz free energy. (See [37] for details of applications). For different introductions of internal variables see [41], [42], [43], [44], [45]. Classical (that is Lagrangian) variational principles for such materials are introduced in [37].

An apparent close relation between the theory of internal variables and statistical theories of continua will not be investigated in this monograph. Lectures of Valanis [44] offer a reasonable introduction to the thermodynamics of continuous media from the internal variables point of view.

Chapter 2

References for Chapter 2

[1] F. John, Partial Differential Equations, Springer Verlag, Berlin and New York, 1971.

[2] S. Lie and F. Engel, Gesammelte Abhandlungen, Teubner, Leipzig, 1922.

[3] L.V. Ovsiannikov, Group-theoretic properties of ordinary differential equations, Acad. Sciences of U.S.S.R., Novosibirsk, 1962.

[4] M. Demiralp and H. Rabitz, Chemical Kinetic functional sensitivity analysis, Journal of Chemical Physics, vol. 75, 1981, p. 1810-1820.

[5] M. Koda, A.H. Dogru and J.H. Seinfeld, Sensitivity of partial differential equations with application to reaction - diffusion processes, J. of Computational Physics, vol. 30, 1979, p. 259 - 282.

[6] A.A. Berezovskii, Lectures on nonlinear problems of mathematical physics, Acad. Sci U.S.S.R., Kiev, 1962.

[7] B. Noble, Complementary Variational principles I, Report # 355, Math. Research Center University of Wisconsin, Madison, Wisc., 1964.

[8] P.D. Robinson, Complementary variational principles, in "Nonlinear Functional Analysis and Applications" (L.B. Rall, Ed.), pp. 507-576, Academic Press, New York, 1971.

[9] M. Kac, Can one hear the shape of a drum? Amer. Math. Monthly, 73, (1966), p. 1-23.

[10] J. Cea, Identification de domaines, Lecture Notes in Computer Science, volume #3, Springer Verlag, Berlin and New York, 1973.

[11] J. Cea, A. Gioan and J. Michael, Queleque resultats sur l'identification de domaines, Calcolo, # 3-4, (1973).

[12] J. Hadamard, Memoire sur le probleme de analyse relatif a l'equilibre des plaques elastiques encastres, 1908 Ouvrés de J. Hadamard, C.N.R.S., Paris, 1968.

[13] N. V. Banichuk, Optimization of the shapes of elastic bodies, Nauka, Moscow, 1980 (in Russian).

[14] N. V. Banichuk, Optimization of the shapes of curved beams, Izv. Akad. Nauk S.S.R., M.T.T. (1975) #6, p. 124-133.

[15] B. Rousselet, Etude de la régularité de valeurs propre par rapport a des deformations bilipschitziennes du domaine geometrique C.R. Acad. Sci., Paris, Ser. A. 283, (1976), p. 507-509.

[16] N.V. Banichuk, Optimization of the shapes of elastic plates subjected to bending, Izv. Akad. Nauk S.S.R., M.T.T., #5, p. 180-188.

[17] E.J. Haug, C.K. Choi and V. Komkov, Design Sensitivity theory for mechanical and structural systems. Academic Press, N.Y., 1986.

[18] K.K. Choi, Shape design sensitivity analysis of displacement and strain constraints, J. Struct. Mech. Vol. 13, #1, (1985) p. 27-41.

[19] K.K. Choi and E.H. Haug, Shape design sensitivity of elastic structures, J. Struct. Mech. Vol. 11 (1983), p. 231-269.

[20] E. Noether, Invariante Variationsprobleme, Nachr. Akad. Wiss. Gottingen, Math. Phys. Kl.II(1918), 235-257.

[21] I.H. Ibrahimov, Invariant variational problems and conservation laws(comments on the theorem of E. Noether), Teoret. Mat. Fiz. 1, No. 3(1969), 350-359.

[22] E.J. Haug, "A Review of Distributed Parameter Structural Optimization Literature," Optimization of Distributed Parameter Structures (E.J. Haug and J. Cea,Ed.), Sijthoff & Noordhoff, Alphen aan den Rijn, Netherlands, 1981, pp. 3-68

[23] J.B. Keller, The shape of the strongest column, Archives of Rational Mechanics and Analysis, Vol. 5 (1960), p. 275-285.

[24] V. Komkov, Application of Invariant Variational Principles to the Optimal Design of a Column, Z.A.M.M., Vol. 61, (1981), p.75-80.

[25] Grässer, H.S.P., A monograph on the general theory of second-order parameter invariant problems in the calculus of variations, Math. Commun. of Univ. of South Africa, Pretoria, 1967.

[26] Eisenhart, L.P., Continuous groups of transformations, Princeton Univ. Press, Princeton, N.J., 1933.

[27] Rund, H., A direct approach to Noether's theorem in the calculus of variations, Utilitas Math. 2,(1972), 205-214.

[28] A. Trautman, Noether's equations and conservation laws, Comm. Math. Phys. 6, (1967), p. 248-261.

[29] J.K. Knowles and E. Sternberg, On a class of conservation laws in linearized, and finite elasticity,Arch. Relational Mech. Anal. 44

(1972), 187.

[30] E. Bessel-Hagen, Über die Enhaltungssatze der Elektrodynamik, Math. Ann. 84 (1921), 258-276.

[31] R.M. Miura, The Korteweg-de Vries equation, Survey of results, S.I.A.M. Revue. 18, #3, (1976), p. 412-459.

[32] V. Komkov, A dual form of Noether's theorem with applications to continuum mechanics, J. Math. Anal. Appl., 75, #1 (1980), p. 251-269.

[33] Anderson, D., Noether's theorem in general mechanics, J. Phys. A., 6, (1973), 209-305.

[34] S. Drobot and A. Rybarski, A variational principle in hydrodynamics, Arch. Rational Mech. Anal. 2, No. 5 (1958), 393-410.

[35] A.M. Arthurs, Complementary Variational Principles, Oxford Univ. Press, 1970.

[36] Nashed, M.Z., 'Differentiability and related properties of non-linear operators: some aspects of the role of differentials in nonlinear functional analysis', in Non-linear Functional Analysis and Applications, edited by L.B. Rall, Academic Press, New York, pp. 103-309.

[37] I. Herrera and M.J. Sewell, Dual extremum principles for non-negative unsymmetric operators, J. Inst. Maths. Applics, 21(1978) p. 95-115.

[38] M.E. Gurtin, Variational principles for linear initial value problems, Quart, Appl. Math., 22, 252, 1964.

[39] K.C. Valanis and V. Komkov, Irreversible thermodynamics from the point of view of internal variable theory (a Lagrangian formu-

lation), Arch. Mech. 32, #1 (1980) p. 33-58.

[40] I. Herrera and J. Bielak, Dual variational principles for diffusion equations, Quart. Appl. Math., 85, 1976.

[41] J. Lubliner, On the structure of rate equations for materials with internal variables, Acta Mechanica, 17, 109, 1973.

[42] K.C. Valanis, Irreversible thermodynamics with internal inertia, principle of stationary total dissipation, Arch. Mech. 24, 5, 948, 1972.

[43] K.C. Valanis, Thermodynamics of internal variables in the presence of internal forces, Report G123-DME-76-0006, Division of Materials Engineering University of Iova.

[44] K.C. Valanis, Twelve lectures in thermodynamics of continuous media, International Institute of Mechanical Sciences, Udine, Italy 1971.

[45] B.D. Coleman and M. Gurtin, Thermodynamics of internal variables, J. Chem. Phys., 47, 597, 1967.

[46] V.E. Stepanov and Iu. A. Iappa, Lie derivative for geometric objects, Vestnik Leningrad. Univ. #22(1978), p. 42-46.

[47] A. Lichnerowicz, Geometrie de Groupes des Transformations, Dunod, Paris, 1958.

[48] E.A. Desloge and R.J. Karch, Noether's theorem in classical mechanics, Amer. J. Phys. 45, #4 (1977) p. 336-339.

[49] E. Tonti, On variational formulation of initial value problems, Ann. Mat. Pure Appl. 95 (1970), p. 331-360.

Chapter 3

Symmetry and Invariance

3.1 An elementary conservation formula.

The invariance conditions related to engineering design practice concern the invariance of the laws of physics, the invariance of imposed conditions, such as boundary conditions or initial conditions that are outside the reach of the designer and the invariance conditions that are chosen by the designer, such as for example constant weight, or constant stress level, and invariance conditions that are not imposed by the designer and arise as a result of a specific design, but may be violated if a different design practice is adopted.

The relation between invariance and symmetry properties has been a subject of considerable research. Let us first consider some conservation laws. A possible definition of a conservation law, or conservation property is the invariance of an operator representing the state, or substate of a system (that is of the state function in the jargon of modern physics) as a function of time only, or of an operator representing some obserable quantity within the system.

The physicists generally insist on more stringent definition, interpreting the statement "conservation law exists", for some observable physical quantity - by postulating that neither the operator defining that quantity nor its expected value vary with time. As an example, one could look at a dynamic system in terms of the relations between final and initial states of such a system. Thus, an operator acting on an initial state produces the state of the system at time t.

The word operator is deliberately vague. Its domain, codomain and range are different in different physical applications. Mathematically unsatisfactory situation is deliberately perpetuated when we associate with the word "operator" its formal properties. For example the operator $\frac{d}{dx}$ mapping $C^{\infty}(\Omega)$ into $C^{\infty}(\Omega)$ functions, $\Omega \subset \mathbb{R}^n$ is a different operator to $\frac{d}{dx}$ mapping distributions of Schwartz (K') into K'. Following the physics and engineering practice we delay the identification of the appropriate spaces, the domain, co-domain and range until the last possible minute, when absence of such information makes the whole discussion mathematically nonsenical

The entire scattering theory can be framed in this type of formalism. An operator S acts on the initial state to produce a finite state of a system. If this operator is linear and finite dimensional (or approximated by a finite dimensional operator) we are at the starting point of the S-matrix theory. In many applications to engineering, to geophysics, or non-destructive testing of materials, one wishes to determine the existence of inhomogeneities, or obstacles to wave propagation. The scattering operator S produces the transition from initial state of the system to the final state. While in each instance experimental results determine the entries of the S-matrix, some of its properties can be deduced a-priori.

If S preserves the probabilities the sum of all probabilities and the sum of final probabilities must be equal to one and it is not hard to show that S is a unitary operator. If the domain of S, i.e. the totality of initial states has countable bases ϕ_i and the expectation of initial state is the same as of the final state, we have

$$\Sigma \,\| S\phi_i \|^2 = 1 \text{ and } \langle S\phi_i, S\phi_i\rangle = \langle \phi_i, \phi_i\rangle$$

thus $S^*S = I$

In this case we have the conservation of the expected value. The more commonly encountered conservation laws arise in reducing some property of the system to the form given in (1.13)-(1.13^a)

(3.1) $$\frac{\partial R}{\partial t} = \frac{\partial Q}{\partial x} .$$

If both R(x,t) and Q(x,t) are integrable on

$-\infty<x<+\infty$, then

$$\frac{d}{dt}\int_{-\infty}^{+\infty} R(x,t)\,dx = -[\ Q(x,t)\ ]_{x\,=\,-\infty}^{x\,=\,+\infty}$$

If we make suitable assumptions (hopefully, based on physical arguments) concerning the behavior of Q(x,t) at infinity, such as: Q(x,t) is time independent at $x = \pm\infty$, or for any value of $t \in [0,\infty]$ $\lim_{x\to+\infty} Q(x,t) - \lim_{x\to-\infty} Q(x,t) = \text{constant}$, we have the conservation law

$\int_{-\infty}^{+\infty} R(x,t)\,dx \equiv \text{constant}$. R(x,t) is sometimes called "the density function". For example, the Burger's equation may be written in the form

$$-\frac{\partial u}{\partial t} = \frac{\partial}{\partial x}[\tfrac{1}{2}u^2 - \nu u_x]$$

where $\lim_{x\to\pm\infty} u\,(x,t) = \text{constant}$, and

$$\lim_{x\to\pm\infty} u_x = 0$$

Thus u(x,t) is the conserved density function and

$\int_{-\infty}^{+\infty} u(x,t)\,dx \equiv \text{constant}$. Returning to structural stability problems we can establish conservation laws by following the Noether approach or by using a version of the (3.1) conservation law.

Example

Let us consider an eigenvalue problem

$Lu = \lambda u$, $\langle u,u\rangle = 1$, where $u \in H_1$ (a Hilbert space)

L is a self-adjoint operator. $L: H_1 \to H_2$, (where H_2 is another Hilbert space) and u, L and λ all depend on a parameter $t \in [0,T)$, where t can be interpreted either as time, or as some design variable, and $T = +\infty$ is not excluded.

Then

$Lu_t + L_t u = \lambda_t u + \lambda u_t$. Here subscript t denotes partial differentiation with respect to t.

The operator L_t, supposedly, may be represented in the form

(3.2^a) $L_t = BL + LA.$

Hence,

$$Lu_t + BLu + LAu = \lambda_t u + \lambda u_t,$$

$$Lu_t + B\lambda u + LAu = \lambda_t u + \lambda u_t,$$

$$\langle Lu_t, u\rangle + \lambda\langle Bu,u\rangle + \langle LAu,u\rangle = \lambda_t\langle u,u\rangle + \lambda\langle u_t,u\rangle$$

or

$$\lambda\langle u_t, u\rangle + \lambda\langle Bu,u\rangle + \lambda\langle Au,u\rangle = \lambda_t\langle u,u\rangle + \lambda\langle u_t,u\rangle$$

and

$$\lambda_t = \lambda\langle (B+A)u,u\rangle$$

Peter Lax considered the case when $B = -A$, i.e. L_t can be written as a commutator

(3.2^b) $L_t = LB - BL.$

Then $\lambda_t \equiv 0$ and λ is an invariant.

Unfortunately, most operators occurring in the structural or mechanical engineering eigenvalue problems do not fit this form. Even if we consider the simplest case of the buckling equation for a beam

$$Ly'' = (Sy'')'' = \lambda y'',$$

where $S = S(h(x))$, $S > 0$, with the operator $L(\cdot) \equiv (\frac{d^2}{dx^2} \cdot S)(\cdot)$, acting on y'', it should be clear that the eigenfunctions of L generally depend on the design and therefore the representation formula (3.2b) cannot be true in the general case. We investigate the special case when such representation formula exists.

Theorem 3.1 A necessary and sufficient condition for (local) independence of the eigenvalue of the beam operator L from the design variable h is the existence of an operator

$$\mathfrak{a} = H_0^2 \rightarrow H_0^2 \quad \text{such that}$$

$$(*) \quad L\mathfrak{a} - \mathfrak{a}L = \left(\frac{S_h}{S}\right) L, \text{ or } L\mathfrak{a} = \mathfrak{a}L + \left(\frac{S_h}{S}\right) \cdot L$$

(recall that $S>0$)

Proof: a) Sufficiency

Suppose such operator $\mathfrak{a}$:

$L_2[0,\ell] \rightarrow H^{-2}[0,\ell]$ exists. Let us denote the operator

$$\left(\frac{S_h}{S} + \mathfrak{a}\right) = B$$

Since $L_h = \dfrac{d^2}{dx^2} S_h = L\left(\dfrac{S_h}{S}\right)$,

we have

$$L_h = L\left(\frac{S_h}{S} + \mathfrak{a}\right) - L\mathfrak{a} = LB - BL$$

which is the condition (3.2^b).

Necessity. Suppose that $L_h = L\left(\frac{S_h}{S}\right)$ can be written in the form $L_h = LB - BL$, or

$$L\left(\frac{S_h}{S}\right) = LB - BL$$

Then $L\left(\dfrac{S_h}{S} - B\right) = -BL$

Denoting

$$\frac{S_h}{S} - B = -\mathfrak{a}$$

we have

$$L\cdot\mathcal{A} = (\frac{S_h}{S} + \mathcal{A})\cdot L$$

or $L\mathcal{A} - \mathcal{A}L = (S_h/S)\cdot L$, as required.

To apply this result we need a lemma that is well known to mechanical and structural engineers but is hard to find in the literature.

Lemma 3.1. Let us assume that a load is independent of the design. A sufficient condition for independence of the bending moment M from any design variable h is the vanishing of the bending moment at the boundary points $x = 0$, $x = \ell$. That is, for any given $y(0)$, $y(\ell)$, we have $M(0) = 0$, $M(\ell) = 0$.

Proof. Let $M(I(x),x)$ denote the function $M(I(x),x) = E\,I(x)\,y''(x) \in L_2[0,\ell]$ (for the current values of $I(x)$). The function $q(x) \in H^{-2}[0,\ell]$ obeys the relation

$$\frac{d^2}{dx^2}\, M(I(x),x) = q(x).$$

$q(x)$ is independent of the design variable $h(x)$.

Since $q_h \equiv 0$, we have

$$\frac{d^2}{dx^2} M_h(I(x),x) = 0, \text{ or}$$

$$\frac{d^2}{dx^2}(EI_h \frac{d^2y}{dx^2}) + \frac{d^2}{dx^2}(EI \frac{d^2}{dx^2}(y_h)) = 0.$$

Thus, $EI_h y'' + EIy_h'' = C_1 + C_2\, x = M_h(x)$. Since at $x = 0$ the moment $M = 0$ independently of h, we have $M_h(0) = 0$ and $C_1 = 0$. Since $M_h = 0$ at $x = \ell$, we have $C_2 = 0$ and $M_h \equiv 0$ for all $x \in [0,\ell]$, as

was to be shown.
Similarly, it can be shown that for a beam cantilevered at one end the bending moment is independent of the design. In fact all statically determinate cases are covered by this statement. This comes as no surprise to any engineer who discovered this fact in an elementary statics course. We combine this statement with the formulas

$$L_h = L\left(\frac{S_h}{S}\right) = LB - BL,\ L^*y = M,\ M_h \equiv 0,$$

$$Ly'' = \lambda y''$$ (The formal adjoint of L is L^* given by

$$L^* = S\cdot\frac{d^2}{dx^2}\ .)$$

Since $L^*\,y = M$, we have $L_h^*\,y + L^*\,y_h = 0$, for a statically determinate case. Then

$$S_h y'' + S y_h'' = 0,$$

$$\left(\frac{S_h}{S}\right) y'' = -y_h'',$$

$$L\left(\frac{S_h}{S}\right) y'' = -Ly_h'',$$

and $L_h y'' = -Ly_h''$.

For local design independence we require

$$L_h = LB - BL\ .$$

Hence

$$(LB-BL)\hat{y}'' = L\hat{y}_h''$$

$$L(B\hat{y}'' - \hat{y}_h'') = BL\hat{y}'' = \lambda B\hat{y}''.$$

Hence, B satisfies the relation.

$$(\ *\ )\quad L B\hat{y}'' - \lambda B\hat{y}'' = L\hat{y}_h'' , \qquad x \in [0, \ell].$$

where $\hat{y}''$ is an eigenfunction of L and λ the corresponding eigenvalue. Thus we have the following theorem.
For a given design S in a statically determinate beam problem, the eigenvalue λ is (locally) design independent if there exists an operator

$B : L_2 \to H^{-2}$ such that (*) is satisfied. We

comment that $S = EI(x)$, that is S is proportional to the moment of inertia of the cross-section about the neutral axis. If we choose the total cross-sectional area $A(x)$ as the parameter, keeping some geometric dependence $I = I(A(x))$, it is not impossible for S_A to vanish. In that case

$y_A = 0$, and λ is an eigenvalue of L with both y''

and By'' being the corresponding eigenfunctions. It suffices to choose the identity operator to represent B to show that λ is locally design independent, that is the value of λ is stationary under small perturbations of S.
Example. Some Invariants of the Korteweg-de Vries (K-de V) equation.
A "standard" form of this equation is
(K-de V) $u_t = 6\, uu_x + u_{xxx}$,

It was already shown that

$$\frac{\partial u}{\partial t} = \frac{\partial}{\partial x}\,(\,3u^2 + u_{xx}), \text{ with } \lim_{|x|\to\infty} u(x,t) = 0$$

implies that $\int_{-\infty}^{+\infty} u(x,t)\,dx$ = constant and u is the

density invariant. This equation may be also rewritten as:

$$(k+1)u^k u_t = (k+1)[6(u^{k+1}u_x) + u^k u_{xxx}] ,$$

or $\frac{\partial}{\partial t} u^{k+1} = \frac{\partial}{\partial x} [6 \frac{(k+1)}{k+2} u^{k+2} + \frac{\partial}{\partial x} (u^k u_{xx}) -$

$(k u^{k-1} u_{xx})$.

But $k(u^{k-1} u_{xx} u_x) = \frac{1}{2}k \frac{\partial}{\partial x} (u^{k-1} u_x^2) -$

$\frac{1}{2}k(k-1)(u^{k-2} u_x^3)$, and so on... Putting k = 1, we derive

$$\int_{-\infty}^{-\infty} u^2 \, dx = \text{constant}.$$

Obviously, we can generate an infinite sequence of independent invariants.

For a discussion of polynomial conservation laws for the K de V equation see Miura [5] and M.V. Novitskiĭ [8].

A simple example.

We consider a one dimensional motion of a mass particle given by the Euler-Lagrange equation

$$(*) \quad \frac{\partial L}{\partial q_i} = \frac{d}{dt}\left(\frac{\partial L}{\partial \dot{q}_i}\right),$$

Let $\int_{q_i=-A}^{+A} \frac{\partial L}{\partial q_i} dq_i = \int_{t_1}^{t_2} \left(\frac{\partial L}{\partial q_i} \dot{q}_i\right) dt = 0$ along a trajectory $\hat{q} = \hat{q}(t)$ satisfying (*)

with $\hat{q}_i(t_1) = -A$, $\hat{q}_i(t_2) = +A$,

Then $\int_{t=t_1}^{t=t_2} \frac{\partial L}{\partial \dot{q}_i} dt$ is an invariant of motion.

3.2 Cartan forms and invariants.

Because of the predominantly engineering audience, this work deliberately avoids proper mathematical introduction to modern differential geometry. A heuristic introduction to the ideas of a tangent space, tangent bundle, cotangent bundle was given in chapters 1 and 2. However, this presentation is sketchy, incomplete and sometimes even misleading if one is not introduced to the theory of differentiable manifolds, exterior products and exterior derivatives. There are several monographs on the market offering an extensive exposition. For example, the five volume treatise of M. Spivak [11], the two volume monograph of Dubrovin, Fomenko and Novikov [12], or the Lovelock and Rund text [13] serve this purpose to a different degree of depth and choice of topics. All offer an excellent introduction to this important and rapidly moving area of mathematics. For a brief introduction to uses of exterior calculus one can study Chapter 7 of Arnol'd's monograph [10], or volume 1 of Thirring's "A course in mathematical physics"[14].

Arnol'd claims that Hamiltonian mechanics cannot be properly understood without a good understanding of differential forms.

Such forms can be introduced recursively. A zero form is a scalar, that is (in general) a real or complex number. For the sake of simplicity let the field of scalars be $\mathbb{R}$ (the real number field). Let ξ_ℓ denote vectors in a linear n-dimensional vector space(over $\mathbb{R}$) that is isomorphic to $\mathbb{R}^n$. Let $\{x_1,x_2,\dots x_n\}$ be coordinates of $\mathbf{V}$. For convenience we could insist that x_i are of unit length, that is normalized. Then any expression

$$w = \sum_{i=1}^{n} a_i x_i,\ a_i \in \mathbb{R} \text{ can be regarded as a 1-form.}$$

The one form may be considered as a functional acting on vectors in $\mathbf{V}$. The value of w at the vector $\xi \in V(\simeq \mathbb{R}^n)$ is given by the number

$$w(\xi) = a_1\langle x_1,\xi\rangle + a_2\langle x_2,\xi\rangle + \dots + a_n\langle x_n,\xi\rangle,$$

where $\langle x_i,\xi\rangle$ denotes the length of the projection of ξ on x_i. Since each unit vector x_i can be regarded as a one-form, we can replace the inner product $\langle x_i, \xi \rangle$ by the evaluation map $x_i(\xi)$.

(Note: $x_i = 0 \cdot x_1 + 0 \cdot x_2 + .. 1 \cdot x_i + 0 \cdot x_{i+1} + \cdots + 0 \cdot x_n$.)

Following the ideas of Hertz we can define forces acting on a mechanical system as one-forms acting on the displacement vectors in $\mathbb{R}^n$.
Clearly, it is immaterial whether we consider one-forms to be forces acting on displacement vectors, or vice versa, as displacements acting on the generalized force vectors. Thus, one-form can be regarded as directed length, or as (signed) vector measure. We define scalar multiplication:

$$(C\omega)(\xi) = C(\omega(\xi))$$

An exterior 2-form ω^2 is a bilinear functional mapping $\mathbb{R}^n \times \mathbb{R}^n \to \mathbb{R}$, that is skew symmetric, i.e.

$$\omega^2(\xi_1, \xi_2) = -\omega^2(\xi_2, \xi_1)$$

and

$$\omega^2(\lambda\xi_1, \xi_2) = \lambda\, \omega^2(\xi_1, \xi_2) \qquad (3.4)$$

$$\omega^2((\xi_1 + \xi_2), \xi_3) = \omega^2(\xi_1, \xi_3) + \omega^2(\xi_2, \xi_3)$$

An example of a 2-form is the exterior product of two one-forms:

$\omega^2 = \omega_1 \wedge \omega_2$, where the bilinear functional $\omega_1 \wedge \omega_2 : \mathbb{R}^n \times \mathbb{R}^n \to \mathbb{R}$ is defined by the formula

$$\omega_1 \wedge \omega_2 (\underset{\sim}{\xi}_1, \underset{\sim}{\xi}_2) = \begin{vmatrix} \omega_1(\xi_1) & \omega_1(\xi_2) \\ \omega_2(\xi_1) & \omega_2(\xi_2) \end{vmatrix} .$$

It is easy to check that all conditions of (3.4) are satisfied for the exterior product $\omega_1 \wedge \omega_2$, and therefore that $\omega_1 \wedge \omega_2$ is a 2-form. The oriented area contained in the parallelogram formed by taking the usual cross product of vectors $\mathbf{v}_1$ and $\mathbf{v}_2 \in \mathbb{R}^3$ is given by $\mathbf{v}_1 \times \mathbf{v}_2 = \{|\mathbf{v}_1| \cdot |\mathbf{v}_2| \sin(\mathbf{v}_1,\mathbf{v}_2)\}\mathbf{k}$, where $\mathbf{k}$ is a unit vector perpendicular to the plane of $\mathbf{v}_1,\mathbf{v}_2$.

The magnitude of the oriented area (positive, or negative) can be computed by assigning an orthogonal coordinate system $\{\mathbf{e}_i\}$ so that the plane spanned by $\mathbf{v}_1,\mathbf{v}_2$ is spanned also by $\mathbf{e}_1$ and $\mathbf{e}_2$. Thus $\mathbf{v}_1 = a_{11}\,\mathbf{e}_1 + a_{12}\,\mathbf{e}_2$, $\mathbf{v}_2 = a_{21}\,\mathbf{e}_1 + a_{22}\,\mathbf{e}_2$. Then the magnitude of the area parallelogram $S = |\mathbf{v}_1 \times \mathbf{v}_2| = \begin{vmatrix} a_{11} & a_{12} \\ a_{22} & a_{22} \end{vmatrix}$.

(See Felix Klein [20].)

On a differentiable manifold M we can identify the infinitesimal area $dx_1\,\mathbf{e}_1 \times dx_2\,\mathbf{e}_2$ with the infinitesimal 2-form $\omega^2 = dx_1 \wedge dx_2$.

K-forms. An exterior k-form ω^k(or form of degree k) is a functional mapping an ordered k-tuple of vectors in a linear vector space $V \simeq \mathbb{R}^n$ into the real numbers, which is k-linear and antisymmetric. That is

$$\omega^k(\xi_1, \xi_2,\ldots,(\alpha\xi_m^{(1)} + \beta\xi_m^{(2)}), \xi_{m+1},\ldots\xi_k)$$

$$= \alpha\omega^k(\xi_1,\ldots\xi_m^{(1)}, \xi_{m+1},\ldots\xi_k)$$

$$+ \beta\omega^k(\xi_1,\ldots\xi_m^{(2)}{}_m, \xi_{m+1}, \ldots\xi_k)$$

and

$$\omega^k(\xi_1, \xi_2,\ldots\xi_k) = (-1)^\beta\, \omega^k\,(\xi_{p1}, \xi_{p2}\ldots\xi_{pk}),$$

where $\xi_{p1}, \xi_{p2} \ldots \xi_{pk}$ is a permutation of the indices 1,2,...k. Then $\beta = 0$ if this permutation is even, and $\beta = 1$ if it is odd.

Lemma. Every k-form on $\mathbf{V} \simeq \mathbf{R}^n$ represents a signed k-volume (that is oriented volume) of a parallelopiped (which could be equal to zero).

Differential forms.

A k-differential form on an n-dimensional (differentiable) manifold is a form:

$$\sum a_{i_1 i_2 \ldots i_k}(\underset{\sim}{x}) dx_{i_1} \wedge dx_{i_2} \ldots \wedge dx_{i_k}$$

where the summation is taken over all repeating indices. The antisymmetric rule still applies. If we interchange the order in a product of two differentials, the sign of the product is altered: $dx_{i_m} \wedge \ldots dx_{i_{m+1}} = -dx_{i_{m+1}} \wedge \ldots \wedge dx_{i_m}$. The form is differentiable if the functions $a_{i_1 \ldots i_k}(x)$ are differentiable. The form has compact support if all these functions have compact support. A product of two differential forms

$$\alpha^k = \sum a_{i_1 \ldots i_k} \, dx_{i_k} \wedge dx_{i_1} \wedge \ldots \wedge dx_{i_k}$$

and $\beta^\ell = \sum b_{j_1 \ldots j_\ell} \, dx_{j_1} \wedge \ldots \wedge dx_{j_\ell}$

is $\alpha\beta^{k+\ell} = \sum a_{i_1 \ldots i_k} b_{j_1 \ldots j_\ell} \, dx_{i_1} \wedge \ldots \wedge dx_{i_k} \wedge dx_{j_1} \wedge \ldots dx_{j_\ell} = -\beta\alpha^{k+\ell}$. $\alpha\beta$ is a multilinear antisymmetric linear functional. It is the zero functional if any two differentials dx_{i_s} are equal.

The exterior derivative of a k-differential form

$$\alpha = \sum a_{i_1 \ldots i_k} dx_{i_1} \wedge \ldots \wedge dx_{i_k}$$

is a (k+1) - form:

$$d\alpha = \sum_{i_1,\ldots,i_k} \left(\sum_{j=1}^{n} \frac{\partial a_{i_1 \ldots i_k}}{\partial x_j} dx_j \right) \wedge dx_{i_1} \wedge \ldots \wedge dx_{i_k}$$

There are no second non-trivial exterior derivatives, because, as can be easily checked by using antisymmetry of differential forms, $d(d\alpha) \equiv 0$.

A form α is called <u>exact</u> if $d\alpha \equiv 0$. Clearly, any form that is the exterior derivative is exact. However the converse statement is false in general. After some trial and error one can come up with examples of exact forms which are not derivatives of other forms.

<u>Contractions</u>.

Let w^k be a k-form and ξ a vector chosen from a vector field on a manifold M. The action of w^k on ξ produces a mapping of w^k into a k-1 form called the contraction of w^k by ξ. This contraction of $w^k(\xi)$ is denoted by $(\xi \lrcorner w^k)$. For example, let k=2, let w^2 be a two-form, ξ a vector field then $\xi \lrcorner w^2$ is a one form, such that for any other vector η, the following equality holds $(\xi \lrcorner w^2)(\eta) = w^2(\xi,\eta) \in R$. Specifically, let x,y be Cartesian coordinates, $dx \wedge dy$ be the 2-form w^2 and $\Phi(x,y)$ a scalar, grad Φ a vector field.

Then $(\text{grad } \Phi \lrcorner dx \wedge dy)(\underset{\sim}{\psi}) = (\nabla\Phi \lrcorner (dx \wedge dy))\underset{\sim}{\psi} = (dx \wedge dy)(\nabla\Phi, \underset{\sim}{\psi})$ for any vector field $\underset{\sim}{\psi}$.

Mentally we can construct an image of the infinitesimal signed area $dx \wedge dy$ (regarded as a 2-form), multiply it by the cosine between the normal to it and the grad Φ direction, (carefully watching orientation) then construct a vector of that magnitude in the direction of grad Φ. That is exactly the contracted one form (that is, the vector field $\underset{\sim}{v} = (\text{grad } \Phi \lrcorner dx \wedge dy)$), that is now regarded as the dual of the vector field $\underset{\sim}{\psi}$. The action of $\underset{\sim}{v}$ on $\underset{\sim}{\psi}$ obeys the usual algebraic laws

of dot product, i.e. the inner product, except for $\pm$ signs. The difference between the more familiar engineering manipulation of tensor products involving the super and subscripts and this formalism is the much greater freedom, independence of the local coordinate atlas and deep connections to differential geometry, such as, for example the immediate availability of a general form of Stoke's theorem in any number of dimensions. In comparing the methods of exterior calculus with the classical Lagrangian and Hamiltonian forms we have to ask which point of view is superior; which point of view allows us to formulate easier the variational or optimization principles for very complex engineering or physics problems? The advantage of Lagrangian mechanics, perhaps combined with Legendre transformation, is the familiarity of this approach. Most graduate students in physics and engineering mechanics are familiar with the classical theories. With some additional training in functional analysis they can make the transition to the modern point of view based on the Lagrangian approach. Conservation laws based on this classical formulation may be derived either from the symmetry of the Lagrangian functional with respect to some variable, or parameter, or - stated in the language of this volume - they can be derived from the property of the group G of automorphisms of the kinematic manifold M and associated Lie subgroups of transformations that leave the action integral J invariant. We accept the commonly used definition that the action integral J is G-invariant, if for any $\underset{\approx}{U} \in G$, $J(y(\underset{\sim}{x}) = J(y'(\underset{\sim}{x}))$, where $y'_i(\underset{\sim}{x}) = \sum_j U_{ij}\, y^j(\underset{\sim}{x})$, with $\delta J(y) = \delta J(y') = 0$.

These are the Noether conservation laws that are homological in nature and can be related to a common mathematical theory known as the de Rham cohomology. This theory lies beyond the scope of this work. It may be found in R. Hermann's book, "vector bundles in mathematical physics"[15], or in his lectures in mathematical physics [16], and is mentioned in the Arnol'd monograph [10], Chapter 7. If the Lie group G is a group of symmetries of the Lagrangian density function, we may refer to conservation of currents, using the electrical analogy. This is the generally used mathematical jargon of topological dynamics. The conservation of the action integral becomes the conservation law for charges.

Other symmetry laws that have nothing to do with the Lagrangian integral relate to the structure of the kinematic manifold, thus depend very strongly on physical constraints assigned to the problem.

As a simple example of this statement consider a motion of a continuous medium (i.e. a flow) such that the equations of motion can be derived from a Lagrangian $\mathcal{L}(t,x,y(t,x)\ y_t(t,x),\ y_x(t,x)\ldots)$, according to Hamilton's principle

$\delta J = \delta \int \mathcal{L}\ dt = 0$. We may investigate the groups of transformations that keep the functional J invariant, thus, return to a restatement of Noether's technique.

However, we may alter, or restrict the kinematic manifold. If $\underset{\sim}{x}$ denotes the local cylindrical coordinates $\{r,z,\theta\}$, $r^2 = x^2+y^2$, $z = z$, $\theta = \arctan\left(\frac{y}{x}\right)$, with a constraint condition that the flow surface is restricted to the cylinder Γ: $r = a$, we acquire new symmetries based on the invariance of any continuous function defined on Γ(not just the Lagrangian) under the transformation $\psi:\{r,z,\theta\} \to \{r,z,(\theta + 2\Pi)\}$.

This symmetry is completely independent of the laws of motion, and appeared when we constrained the kinematic manifold to a cylinder. Such symmetry laws are homotopic in nature and reflect the topological structure of the manifold. The group of symmetries that we introduce here is "the gauge group": $\{r,z,i\theta\} \to \{r,z,i(\theta \pm 2\Pi)\}$.

3.3 The Cartan forms.

The Cartan forms were the outgrow of a course that Cartan taught in rigid body mechanics. After one establishes a kinematic manifold $\mathcal{M}$, that is the space of all admissible configurations of a mechanical system with n-degrees of freedom, the motions on that system are represented by velocity vectors that are tangent to the configuration manifold. Thus, we construct a new manifold consisting of tangent spaces to $\mathcal{M}$ at all points $x \in \mathcal{M}$. The manifold $\bigcup_{x\in\mathcal{M}} TM_x$ is a manifold of dimension 2n, denoted by TM and called the tangent bundle of $\mathcal{M}$. We associate with each vector $\xi \in TM_x$ the point $x \in \mathcal{M}$. The map Π: $TM \to \mathcal{M}$, $\xi \in TM_x \to x(\in \mathcal{M})$ is called the natural projection. The inverse image of $\underset{\sim}{x} \in \mathcal{M}$, that is $\Pi^{-1}(\underset{\sim}{x})$ is called the fiber of the tangent bundle over $\underset{\sim}{x}$. (See Steenrod [17] for a general definition.) The motion of a system with n-degrees of freedom evolves on the Cartesian product $TM \times \mathbb{R}_+$ where $\mathbb{R}_+$ is the configuration set of a real parameter, mentally indentified with time. (But any other parameter may be used .) At each time t we identify the position and velocity on TM by specifying $(\underset{\sim}{x}, \underset{\sim}{v}) \in TM$, and by specifying the evolution of the system, that is the points of $TM \times \mathbb{R}_+$, $p = (\underset{\sim}{x}, \underset{\sim}{v}, t) = (x^1, x^2, \ldots, x^n, v^1, v^2, \ldots \ldots v^n, t)$. There is no reason why more than one parameter could not be used, but for the sake of simplicity let us persist with a single parameter. If t is the time and v is the velocity vector, the obvious choice of one forms for a Lagrangian system is:

$$\begin{cases} dx^i - v^i\,dt & \text{(a)} \\ dv^i - a^i\,dt & \text{(b)} \\ dt & \text{(c)} \end{cases}$$

$i = 1,2,\ldots n.$

The 2n+1 forms (a), (b), (c) can serve as basis for one forms on TM x $\mathbb{R}_+$. We introduce a vector field $\underset{\sim}{F}$ which annihilates the forms (a) and (b).

$$(F,(dx^i - v^i\, dt)) = 0 = (F,(dv^i - a^i\, dt)),$$

where F is an element of TM x $\mathbb{R}_+$. The integral curves of F generate the constants of motion for the Lagrangian system.

A basic difference in language and notation should be noted. We do not speak of finding a solution to the system of Euler Lagrange equations on a kinematic manifold. Instead we are looking for a vector field on TM x $\mathbb{R}$ that annihilates the Euler Lagrange form and makes compatible a vector field $\underset{\sim}{v}$ with the direction of a tangent vector to the kinematic manifold $\mathcal{M}$, and acceleration $\underset{\sim}{a}$ compatible with the cotangent vector bundle at each point $x \in \mathcal{M}$. The Lagrangian function satisfies the following relation on the manifold TM x $\mathbb{R}_+$.

$$(3.6)\quad \frac{\partial L}{\partial x^i} - \frac{\partial}{\partial t}\left(\frac{\partial L}{\partial v^i}\right) = g_{ij}a^j + \frac{\partial}{\partial x^j}\left(\frac{\partial L}{\partial v^i}\right) v^j ,$$

where $g_{ij} = \dfrac{\partial^2 L}{\partial v^i \partial v^j}$ is the kinematic metric, if the Lagrangian is the kinetic energy, that is

$\mathcal{T} = \mathcal{L} = \sum g_{ij}\dot{x}^i\dot{x}^j$. If we identify the vector field $\underset{\sim}{v}$ with the tangent vectors to Lagrangian kinematic manifold and $\underset{\sim}{a}$ with integral curves in TM x $\mathbb{R}_+$, we imply that the vector field F annihilates (a) and (b).

If, in fact, as the system evolves, $v^i = dx^i/dt$, $a^i = \dfrac{dv^i}{dt}$ along F, then we have the relation

$$\frac{d}{dt}\left(\frac{\partial L}{\partial v^i}\right) = \frac{d}{dt}\left(\frac{\partial L}{\partial (\dot{x}^i)}\Big\downarrow_F\right) = \left\{\frac{\partial}{\partial t}\left(\frac{\partial L}{\partial \dot{x}^i}\right) + \right.$$

$$\left[\frac{\partial}{\partial x^j}\left(\frac{\partial L}{\partial \dot{x}^i}\right)\right]\frac{dx^j}{dt} + \left[\frac{\partial}{\partial \dot{x}^j}\left(\frac{\partial L}{\partial \dot{x}^i}\right)\right]\frac{d\dot{x}^j}{dt}\ \}\Big|_{\mathbf{F}}$$

Thus along **F**, we have the vanishing of the Euler-Lagrange functions

$$(3.7)\quad \left(\frac{\partial L}{\partial x^i} - \frac{d}{dt}\left(\frac{\partial L}{\partial \dot{x}^i}\right)\right),\ i = 1,2,\ldots n.$$

The vector field **F** annihilates any form of the type $f_{1_i}(\tilde{x},\tilde{v},t)(dx^i - v^i\,dt)$ +

$$f_{2_i}(\tilde{x},\ \tilde{v},\ t)\cdot(dv^i - a^i\,dt).$$

We make use of this property by identifying $f_{1_i}(\tilde{x},\mathbf{v},t)$ with the second derivative:

$$\frac{\partial^2 L}{\partial x^i\,\partial v^j},\ \text{and } f_{2_i} \text{ with } g_{ij} = \frac{\partial^2 L}{\partial v^i \partial v^j}.$$

Then **F** annihilates the two-form

$$(3.8)\quad \left[\frac{\partial^2 L}{\partial x^i \partial v^j}(dx^i - v^i\,dt) + g_{ij}(dv^i - a^i dt)\right] \wedge (dx^j - v^j\,dt)$$

(See M. Crampin [9] for derivation of (3.8).)

The two-form(3.8) turns out to be the exterior derivative of a simple one-form

$$(3.9)\quad \Theta = L\,dt + \frac{\partial L}{\partial v^i}(dx^i - v^i dt)$$

which is the basic Cartan form for a Lagrangian system (see [9]). It can be checked that

$$(3.10)\quad F \lrcorner\ d\Theta = 0.$$

(Note that $d(dt) \equiv 0$). Thus, to each density Lagrangian Cartan assigns a family of one-forms Θ, that are annihilated by contraction (action on Θ) a vector field $\mathbf{F} \in TM \times \mathbf{R}$.

We say that $\mathbf{F}$ is the characteristic field of $d\Theta$, (or for $d\Theta$). While $\mathbf{F}$ annihilates the forms $(dx^i - v^i\,dt)$, $(dv^i - a^i\,dt)$, (but not dt) Θ depends only on $(dx^i - v^i\,dt)$, and dt, and does not depend on the other basis of $TM \times \mathbf{R}$, that is on $(dv^i - a^i\,dt)$. Therefore Θ lies on a lower dimensional manifold TM_v. In the relation (3.10) $F \lrcorner\, d\Theta = 0$ we notice that altering Θ by some exact differential form on TM_v does not affect this relation. Suppose that $f = f(x,t)$ is a scalar.

Then $df = \frac{\partial}{\partial t}\,dt + \frac{\partial f}{\partial \underset{\sim}{x}}\,d\underset{\sim}{x} =$

$$\frac{\partial f}{\partial t}\,dt + \frac{\partial f}{\partial x^i}\,dx^i = \frac{\partial f}{\partial t}\,dt + \frac{\partial f}{\partial x^i}\,(dx^i - v^i\,dt)$$

$$+ \frac{\partial f}{\partial x^i}\,v^i\,dt = \left(\frac{\partial f}{\partial t} + \frac{\partial f}{\partial x^i}\,v^i\right)dt + \frac{\partial f}{\partial x^i}(dx^i - v^i\,dt).$$

Let $\hat{\Theta} = \Theta + df$.

Then $d\hat{\Theta} = d\Theta$, and $F \lrcorner\, d\hat{\Theta} = 0$.

It follows that Cartan forms annihilated by the vector field $\mathbf{F}$ are unique up to an exact form (think of it as a divergence) on TM_v.

3.4 Some applications of Cartan's exterior calculus to continuum mechanics.

Most fundamental laws of continuum mechanics are expressed by equations of equilibrium, equations of compatibility, or of continuity and constitutive equations of physical substances.

Several authors have made claims that discoveries of invariants and of hidden symmetries are absolutely essential to efficiency of computational techniques that have been particularly troublesome in real-life problems encountered in continuum mechanics. Geophysics, meteorology, oceanography, problems in efficient design of structures or mechanisms require enormous amounts of computing time. The author of this work believes that advances in group theoretic methods and in either more classical tensor or exterior calculus methods will eventually become important to the numerical analysts in all of these fields.

In Lagrangian systems we associate the vector field $\underset{\sim}{W}$ with the generalized force field if x^i denote generalized displacements. If $L(x^i, u^j(x), \partial_k u^j(\underset{\sim}{x}))$ is the Lagrangian density function, we denote by,

$$W_i = \frac{\partial L}{\partial u^i}, \quad W_i^k = \frac{\partial L}{\partial(\partial_k u^i)}$$

computed along a vector field

$$\mathbb{F} : \underset{\sim}{u} = \hat{\underset{\sim}{u}}\,(\underset{\sim}{x}) \text{ on the manifold TM.}$$

First order differential equations are replaced by one-forms

$$\omega^i = du^i - \phi^i_j \, dx^j, \; i = 1,2,\ldots n .$$

Then $(\mathbb{F}, \omega^i) = 0$ if and only if

$$\phi^i_j(\underset{\sim}{x}) = \frac{\partial \hat{u}^i}{\partial x^j} \text{ along the vector field } \mathbb{F}.$$

$$\mathbb{F} : u^i = \hat{u}^i(x^1, x^2, \ldots x^n), \; i = 1,2,\ldots n.$$

The equations of continuity and of equilibrium may be generally reduced to n-th order differential forms:

$$Q_i = W_i \, d(vol) + d(\omega^j_i) \;\; (\partial_j \lrcorner \, d(vol)),$$

where $d(vol) = dx^1 \wedge dx^2 \ldots \wedge dx^n$

represents the infinitesimal (signed) n-dimensional

volume, $\partial_j \equiv \frac{\partial}{\partial x^j}$, so that

$$\partial_j \lrcorner\, \partial v \wedge dx^k = (-1)^p\, \delta_j^k\, (d\ vol)$$

where δ_i^k is the Kronecker's delta

$$\delta_i^k \begin{cases} = 1 \text{ if } k = i \\ = 0 \text{ if } k \neq i\ . \end{cases}$$

p is the permutation order for interchange of the i^{-th} and k^{-th} index.

For example, by following the Saint Venant's assumptions, we can model the torsional vibrations of a rod by the linear differential equation

$$(3.11) \qquad \frac{\partial}{\partial x}\left(G_p \frac{\partial \theta}{\partial x}\right) - \rho I_p \frac{\partial^2 \theta}{\partial t^2} = 0\ ,$$

where $G_p \frac{\partial \theta}{\partial x}$ is the torque applied to the rod.

This is really Newton's second law, or the equilibrium equation for this system.

We recognize that the equation (3.11) is the Euler-Lagrange equation

$$\frac{\partial}{\partial t}\left(\frac{\partial L}{\partial \hat{\theta}_t}\right) - \frac{\partial}{\partial x}\left(\frac{\partial L}{\partial \hat{\theta}_x}\right) = 0.$$

where $L = \frac{1}{2}\left[\rho\, I_p (\theta_t)^2 - G_p (\theta_x)^2\right]$.

We replace the system (3.6) by requirement that the differential form

$$[(L_1^1\ (x,t,\theta,\hat{\theta})\ dx + L_2^1\ (x,t,\theta,\hat{\theta})\ dt)]\ dx \wedge dt$$

is annihilated by **F**.

Hence, $L_1^1 = \frac{\partial L}{\partial(\partial_t \hat{\theta})}$, $L_2^1 = \frac{\partial L}{\partial(\partial_x \hat{\theta})}$.

In problems of fluid flow we encounter various versions of the Navier-Stokes equations featuring the term $\underset{\sim}{\mathbf{v}} \cdot \text{grad}\, \underset{\sim}{\mathbf{v}}$ where $\underset{\sim}{\mathbf{v}}$ is the fluid's velocity. Let V be a vector field

$$V(\underset{\sim}{x},\underset{\sim}{u}, \underset{\sim}{\phi}) = v^i(x,\underset{\sim}{u},\underset{\sim}{\phi}) \frac{\partial}{\partial x^i} + v^j(\underset{\sim}{x},u,\underset{\sim}{\phi}) \frac{\partial}{\partial u^j} +$$

$$v^k_\ell (x,u,\phi) \frac{\partial}{\partial \phi^\ell_k} .$$

We consider the contact 1-forms

$$(\omega^1)^i = du^i - \phi^i_k \, dx^k$$

$i = 1,2,\ldots n$, and the n-forms

$$Q_i = W_i \, d(vol) + d(W^j_i \wedge (\partial_j \lrcorner \, d(vol))$$

$i,j = 1,2,\ldots,n$.

These are annihilated by a vector field $\mathbf{F}$ on the manifold TM. Of particular interest are similarity solutions obeying the constraints

$$v^i(x,u(x)) = v^j(x,u(x)) \frac{\partial u^i}{\partial x^j}$$

corresponding to shift along the vector field

$$\mathbf{V} = \sum v^i \frac{\partial}{\partial x^i} .$$

Example.

A one-dimension gas flow with a gravity force in the x-direction satisfies a system of equations

$$(3.12) \qquad \begin{aligned} u_t &= -p_x \\ v_t &= u_x \\ p_t &= -\mu(v,p)u_x \\ p_x &= a(t,x,u,v,p)u_x + f(t,x,u,v,p)v_x \\ &+ f(t,x,u,v,p) = g. \end{aligned}$$

The system admits a one-parameter group of transformation with a vector field

$$\mathbf{V} = \tau \frac{\partial}{\partial t} + \xi \frac{\partial}{\partial x} + \eta^1 \frac{\partial}{\partial u} + \eta^2 \frac{\partial}{\partial v} +$$

$$\eta^3 \frac{\partial}{\partial p} + \phi^i_j \frac{\partial}{\partial p^i_j} .$$

The system (3.7) may be rewritten as a differential system

$$d(udx + pdt) = 0$$

$$u_t - g = 0$$

$$du - u_x\, dx + g\, dt = 0$$

$$dv - u_x\, dt - v_x\, dx$$

$$dp + \mu\, u_x\, dt - g\, dx = 0 .$$

It can be shown that a necessary condition for this system to admit a vector field $\mathbf{V}$ consistent with a one parameter group of contact transformations is $f(t,x,u,v,p) \equiv 0$ along any integral curve of this system, and that either $\mu \equiv$ constant or else $\mu_v = \mu \cdot \mu_p$.

This is one of the results obtained by the Siberian group on gas dynamics (Krendelev, Ianenko, Talyshev, etc.)

Elastic, or elastoplastic torsion of shafts

In Saint-Venants model of pure torsion only two stress components τ_{xz} , τ_{yz} are not equal to zero in the x-y plane perpendicular to the axis of the shaft which coincides with the z-axis.

The torque is constant and is equal to M. The equilibrium equation is the Saint-Venant's equation

$$(3.13) \qquad \frac{\partial \tau_{xz}}{\partial x} + \frac{\partial \tau_{yz}}{\partial y} = 0 \text{ in } \Omega \subset \mathbb{R}^2,$$

or if we introduce Prandtl stress function $\Phi(x,y)$ such that equilibrium equations become trivial:

$$- \frac{\partial \Phi}{\partial x} = \tau_{yz}, \quad + \frac{\partial \Phi}{\partial y} = \tau_{xz},$$

and , let ω^1 **grad** $\phi = f^1 = \frac{\partial \Phi}{\partial x} dx + \frac{\partial \Phi}{\partial y} dy$,

where ω^1 is the 1-form $\Phi\, dx + \Phi\, dy$,

then $f^1 = - \tau_{yz}\, dx + \tau_{xz}\, dy$, and

the equilibrium equation becomes the identity

$$df^1 = 0.$$

In the (r,θ) polar coordinate system

$f^1 = \omega^1$ **grad** $\Phi = \Phi h_1\, dr + \Phi h_2\, d\theta$, where

$h_1 = 1$, $h_2 = r$.

h_i are the Lammé scalars for orthogonal$\{q_i\}$, $(i=1,2)$ coordinate system in which the metric is given by

$ds^2 = \sum_i h_i^2\, dq_i^2)$. We compute:

$\text{grad}\ \Phi = \frac{\partial \Phi}{\partial r} + \frac{1}{r}\frac{\partial \Phi}{\partial \theta}$. Again we have $df^1 = 0$.

<u>Note</u>

The existence of a plastic region implies existence of a region $\Omega_p \subset \Omega$ in which a condition of plastic flow is satisfied. If one assumes that the Huber-von Mises criterion applies, this is reduced to the equality

$$K^2 \equiv (\tau_{xz}^2 + \tau_{yz}^2) \qquad \varepsilon\ \Omega_p, \text{ or}$$

$$|\text{grad}\ \phi|^2 \equiv K^2 \qquad \{x,y\} \in \Omega_p$$

with $\phi \equiv 0$ on $\partial\Omega$.

$$\phi \equiv \text{const} = A_p \text{ on } \partial\Omega_p$$

$$\phi \in H_0^1\ (\Omega).$$

<u>The Lie derivatives in Cartan's form</u>

For a one form ϕ we define the Lie derivative of ϕ in the direction of a vector field X to be

$$X \lrcorner\, d\phi + d\,(X,\phi) = L_X\ \phi.$$

Let us suppose that $L_X\ \phi = df$.

Then $L_X\ \phi - d(x,\phi) = df - d(x,\phi)$

$$= d(\ f - (X,\phi))$$

But $L_X\phi - d(x,\phi) = X \lrcorner\, d\ \phi$.

Therefore $d(X \lrcorner\, d\phi) \equiv 0$.

A more conventional presentation of Lie derivatives in the setting of continuum mechanics can be given as follows. Let x^i be a coordinate system on a differentiable manifold $\mathbf{V}_n$. The partial derivatives $\frac{\partial f}{\partial x}$ form basis of a vector space T_x (see A. Lichnerowicz [21]).

Since $df = \sum \frac{\partial f}{\partial x^i}\ dx^i$ the vector $\{dx^i\}$ forms basis of the space T_x dual to $^*T_x$ with basis $\frac{\partial}{\partial x^i}$.

The accepted notation that uses the summation convention is

$$df = \partial_i f\ dx^i.$$

We refer (using the language of tensor mechanics) to the elements of $^*T_x$ as covariant vectors and to T_x as contravariant vectors.

As we change coordinate systems

$x^{i'} = f^{i'}(x^i)$ we obtain

$$dx^{i'} = \frac{\partial f^{i'}}{\partial x^j} dx^j = A^{i'}_j dx^j .$$

Thus, we can introduce arbitrary elements ω^α of $^*T_x$ such that

$$\omega^\alpha = a^\alpha_i dx^i .$$

Dual basis are defined by the relations

$$\langle dx^i, \Pi_j \rangle = \delta^i_j, \quad \langle \omega^\alpha, e_\beta \rangle = \delta^\alpha_\beta ,$$

where δ^i_j is the Kronecker's delta symbol

$$\delta^i_j \begin{cases} = 1 \text{ if } i = j \\ = 0 \text{ if } i \neq j. \end{cases}$$

The linear connection A^i_j satisfies

$$\partial_i A^\mu_\nu - \partial_\nu A^\mu_i = 0$$

All transformations corresponding to changes of basis form a group:

$$\mathbf{e}_{i'} = A^j_{i'} \mathbf{e}_j, \qquad \omega^{i'} = A^{i'} \omega^j .$$

Covariant derivative ∇ has the following properties: a) It is a linear operator.

b) Application of ∇ to a scalar function f transforms f into a covariant vector ∇f which may be also denoted by *df.

c) It raises the rank of any tensor by one contravariant index (in the usual engineering interpretation of "tensors")

d) It obeys the Leibniz algebraic law of derivations.

$$\nabla(c_\alpha \omega^\alpha) = (\nabla c_\alpha) \cdot \omega^\alpha + c_\alpha (\nabla \omega^\alpha)$$

with

$$\nabla \omega^\alpha = \Gamma^\alpha_{\beta_\gamma} \omega^\gamma \otimes \omega^\beta$$

Γ^α_β is the linear connection known as the Christoffel symbol.

It satisfies the conjugation law

$$\Gamma^\alpha_{\beta\gamma} = \overset{*}{-}\Gamma^\alpha_{\gamma\beta} .$$

We assert that $\nabla\omega^\alpha$ is a tensor of rank two. The concept of a deformation or of a moving coordinate system is closely related to the concept of an infinitesimal generator of a transformation and the corresponding Lie derivative.

A transformation assigning to a point $\{x^i\} = p$, the point $p' = \{x^i + \xi^i \delta t\}$ can be regarded as an infinitesimal transformation in which $\underset{\sim}{\xi}$ is the infinitesimal generator of a semigroup, or as a process of dragging of a coordinate system through an infinitesimal vector

$$dp = \langle dx^i, \Pi_i \rangle = \xi^i . \mathbf{e}_i \ \delta t.$$

The <u>natural</u> coordinate system $\{\mathbf{e}_i\}$ is defined in [19] by V.E. Stepanov and Iu. A. Iappa.

<u>Another definition of the Lie derivative.</u>

The Lie derivative of a vector $\underset{\sim}{u}$ in the direction of a vector η can be defined as the commutator

$$L_\eta \underset{\sim}{u} = \langle \eta, \nabla \underset{\sim}{u} \rangle - \langle \underset{\sim}{u}, \nabla \eta \rangle$$

The Lie derivative of a scalar function f(x) is defined by the rule

$$L_\eta f = \eta^i \cdot \partial_i f.$$

Hence, it is the directional derivative, that is a projection of gradient of f in the direction of η. For a vector $\underset{\sim}{u} = c^i \mathbf{e}_i$, where $\mathbf{e}_i$ are basis, the Lie derivative is defined by Leibniz rule

$$L_\eta \underset{\sim}{u} = L_\eta (c^i \mathbf{e}_i) = (L_\eta c^i)\, \mathbf{e}_i + c^i (L_\eta \mathbf{e}_i)$$

$$= (\eta^i \partial_i c^j)\, \mathbf{e}_j + c^i (L_\eta \mathbf{e}_i)$$

We recall that

$$L_\eta \mathbf{e}_i = \eta^j \Gamma^k_j \mathbf{e}_k - \nabla_i \eta^j \mathbf{e}_j .$$

Let u be a vector expressed in a moving coordinate systẽm. Let $\mathbf{e}_i$ be local basis.

$$\mathbf{e}_i(\underset{\sim}{x} + d\underset{\sim}{x}) = \mathbf{e}_i(\underset{\sim}{x}) + \Gamma^j_{ik} \eta^k \mathbf{e}_j \cdot \delta t \quad .$$

Lie derivatives are defined by

$$L_\eta u^i \delta t = u^i(x+dx) - \tilde{u}^i(x) ,$$

where $\tilde{u}^i$ is the Lagrangian representation of a vector dragged along a moving coordinate system in the direction η.

$$\tilde{u}^i(\underset{\sim}{x}+d\underset{\sim}{x}) = u^i(\underset{\sim}{x}) + \Omega^i_j u^j(\underset{\sim}{x})\, \delta t ,$$

with $\Omega^i_j = \nabla_j \eta^i - \Gamma^i_{jk} \eta^k$.

3. Introduction to engineering design applications.

The most important engineering activity is design of engineering systems and subsequent control and modifications of design. The behavior of a mechanical system can be represented as a mapping of a differentiable manifold into itself. This modern point of view is pursued in monographs of V.I. Arnol'd [10], W. Thirring [14], R. Abraham and J. Marsden [22]. This approach was originated in a remarkable theory established by E.Cartan, which contributed significantly to new understanding of mechanics and physics only in the last twenty years. The names of Arnol'd, Moser, Smale, Novikov, Souriau, Thirring, Sinai come to mind in connection with this point of view. Continuum mechanics applications of this theory have been slow and infrequent. The exterior algebraic forms have not displaced the commonly accepted engineering (formal) use of tensor calculus. Wave mechanics have been studied on symplectic phase space manifolds, but in general the mechanical or structural problems from the "real world" continue to be studied by established techniques. In recent years spectacular progress has been made by using functional analytic techniques which were developed in the theory of partial differential equations. More specifically, Sobolëv space techniques have been applied to classify mechanical designs and to estimate changes in the performance of structural or mechanical systems. Of course, one could regard the work performed by a system of applied loads either as a one-form acting on the displacements (which are fibers of the Kinematic manifold) or as an inner product in a Sobolëv space.

A second look at Hilbert space arguments results in some doubts about the modelling of physical systems. The careful distinction between displacements and velocities (points on the manifold $\mathcal{M}$ and the tangent manifold TM) and forces (which are points in the cotangent manifold) may be completely obliterated in the Hilbert space setting. One could get around this objection by assigning either a more general topological structure or by keeping track - which functions are in the original Hilbert space H (generalized displacements) and which are in its topological dual H*. (generalized forces). The Sobolëv space structure is convenient for modelling which avoids this difficulty. Another objection which seriously affects irreversible processes has to do with causality. Nobody really studies Hilbert spaces per se. One generally studies transformations, or mappings of Hilbert spaces, i.e. properties of operators on Hilbert spaces. In modelling of dynamic phenomena some authors observed that causality (or anticausality) can be assigned to operators in a Hilbert space H by either forming a Cartesian product of H with a well-ordered set $\tilde{\Gamma}$ (which is generally a homomorph of the real line) or by assigning a group structure G to all points of H. See R. Saeks [23]). This approach is closely related to the existence of a semi-group of operators in evolutionary systems, i.e. with the Hille-Phillips approach [26], which is regarded as a standard technique in a class of partial differential equations. While use of group-theoretic methods has been very limited in design problems ([24] [25] are exceptional examples) it has now reached the status of an acceptable tool in control theory including the control of systems with distributed parameters. However, it is only a question of time before the Lie group techniques are used to generate improvements in engineering design and are incorporated into numerical techniques such as gradient projection which are now heavily oriented towards functional analysis and approximation theory.

1. Preliminary definitions

1. Let $\mathcal{M}$ be a smooth n-dimensional manifold consisting of admissible dynamic configurations for a mechanical system. A theorem of Whitney states that $\mathcal{M}$ may be embedded in a Euclidean space. The tangent bundle of $\mathcal{M}$ (which is the union of tangent spaces $TM_{\tilde{p}}$, $\tilde{p} \in \mathcal{M}$) is denoted by TM. Local coordinate system q forms an atlas over $\mathcal{M}$. The admissible kinematic configurations are determined by constraints

(1.1) $\eta_i(\tilde{q}) = 0, \; i = 1,2,\ldots T.$

A design of the mechanical system consists of specifying a vector $\tilde{S}$ in a compact set $S \subset \mathbb{R}^m$.

The set S is called the set of admissible **designs.** For a fixed value $\tilde{\hat{s}} \in S$ a map $T(t)$: $\mathbb{R}_+ \to M$ is called the motion of the mechanical system. T(t) is assumed to be a differentiable map. $\dot{T}(t) \in TM$ is the velocity map. Assignment of a point $q_0 \in \mathcal{M}$ to T(0) specifies the initial state (or initial conditions) to the motion of a mechanical system. $T(0)$ is the initial velocity. We introduce a class of smooth maps ψ_α: $\mathbb{M} \subset TM \to \mathbb{R}$ which are also elements of a Hilbert space H_1 and a linear transformation A: $H_1 \to H_2$ where H_2 is a Hilbert space. In mechanical problems the spaces H_1, H_2 are generally Sobolëv spaces. For each Fiber which is an admissible motion $\hat{q}(t) = \hat{q}(q_0,t) \in \mathbb{M}$, with $\hat{q}(q_0,0) = q_0$, and corresponding fiber $\dot{\hat{q}} \in TM$, we define for each $\underset{\sim}{s} \in S$ the map $\psi_t\ (\hat{q},\dot{\hat{q}})$, $t \in [0,\infty)$ (or $t \in [0,t_1]$): $\mathbb{M} \times \mathbb{R}_+ \to \mathbb{R}^\ell$, and a set of physical laws and constitutive equations of the form:

$$A\Psi_i = f_i(t) \in H_2(M),\ i = 1,2,\ldots\ell;$$

In specific problems A has some additional properties which need to be specified, such as symmetry, or self-adjoint property, positiveness, etc.

The performance of the system is judged by the value of a map Φ: $\mathbb{M} \to \mathbb{R}$. Specifically the map Φ assigns a real number to each motion of the mechanical system, i.e. $\Phi(\hat{q}(t)) \in \mathbb{R}$ determines the performance of the mechanical system for the motion $\hat{q}(t)$. The equations **(1.2)** shall be called the state equations. The weak form of (1.2) is a map $H_2 \times H_2 \to \mathbb{R}$

$$(1.3)\quad \langle\Psi_i, A^*\Lambda_i\rangle_{H_2} - \langle f_i, \Lambda_i\rangle_{H_2} = 0,$$

where for the present time Λ_i is considered to be any element of $H_2(M)$.

2. Some Examples

Let us offer at this stage some low-dimensional examples of kinematic manifolds.

2a) Let us consider a motion of a single particle in $\mathbb{R}^3$ if no constraints are placed on admissible location of the particle, then $\mathbb{M} = \mathbb{R}^3$.

2b) The motion of a double pendulum

The position of the system is uniquely determined by specifying $\{\theta_1, \theta_2\}$, with $\theta_i \pm 2n\pi$ identified with θ_i for any integer n. $\mathcal{M}$ is a torus (a Cartesian product of two circles).
2c) Rotation of a rigid body about a fixed point. The position of the body is uniquely determined by Euler's angles ϕ, θ, Ψ. $\mathcal{M}$ is the Cartesian product of 3 circles embeded in $\mathbb{R}^4$.
2d) The conservative motion of a gyroscope. $\mathcal{M}$ is the Cartesian product of a circle and a cone in $\mathbb{R}^4$.
2e) Elastic pendulum in coordinates r, θ restricted to a plane.
$\mathcal{M}$ is a semi-cylinder (Cartesian product of a circle with a half ray). Again, we identify θ with $\theta \pm 2n\pi$.
2f) Planetary gear system. $\mathcal{M}$ is a spiral located on a torus.
2g) One of the shafts of the planetary gear system above is regarded as a torsion spring, the other is approximately rigid. $\mathcal{M}$ is the Cartesian product of the curve in (2f) with a circle.

As one can see from these examples the Kinematic manifolds of even simple mechanical systems could have an interesting topological structure. For example, if in a gear transmission design the transmission ratio is irrational (substitute a friction drive instead of gears in (2f) $\mathcal{M}$ could be a Cartesian product of a line with a curve, which fails to be locally connected. The design vector s is easily identified in each example. Vector $\underset{\sim}{s}$ in (2a) is the mass of particle; in (2b) it is the vector $\{\ell_1, \ell_2\}$ of the lengths of the penduli; in (2c) $\underset{\sim}{s}$ is identified with the values of principal moments of inertia $\{I_{xx}, I_{yy}, I_{zz}\}$, in (2e) it is the ordered pair $\{k,m\}$, etc.

<u>A general discussion of the effects design changes on the structure of kinematic manifold.</u>

The "usual" problems discussed in books on mechanics have certain traditional setting that is suitable to educational activities but not very realistic from the point of view of practical applications. The reasons are familiar to any engineer who worked on the analysis and synthesis of electrical networks, structures or mechanisms. In classical or applied mechanics theory one presumes that the design of the system is fixed. The constraints imposed on it are either geometric or physical (such as constitutive equations, and the general laws of physics). Some physical laws may be replaced by the variational statements, thus perhaps converting a complete free motion to one constrained to a geodesic path. In "real" engineering problems design changes are always considered. The possible "small" changes in the design parameter shift a geodesic curve representing the physical behavior to another geodesic. This change may be continuous and smooth, or it may be badly discontinuous or at least non-differentiable. These problems are closely tied to the sensitivity of solutions (i.e. flows on the manifold) to changes in design parameters.

A more common engineering approach is to consider the variational problem of minimizing the cost functional with the equations of state and other equations of physics regarded as constraints. In effect this is a version of the Bolza problem of the classical calculus of variations complicated by changes in the design parameters or in design functions. This version of the Bolza problem may be fairly straightforward, or may be surprisingly difficult, depending on the nature of the constraints, where the differential equations (or integro-differential equations) describing the laws of physics are regarded as constraints. Before proceeding with this rather difficult problem we need to briefly review the differences between Lagrangian and non-Lagrangian equations of state. First, let us offer some easy examples.

Examples of physical laws.

Let us consider the physical example (2e) of an elastic pendulum. We define the functional

$$\Psi_1 = H = \tfrac{1}{2}\{M(r\,\dot{\theta})^2 + m\dot{r}^2\} + \tfrac{1}{2}(Kr^2) + mg\,(h_0 - r\sin\theta),$$ with h_0 chosen arbitrarily.

We postulate the absence of dissipation by stating the physical law
$\frac{d}{dt}(H) = 0$ (in absence of external forces) along any trajectory of the system.

Here $A \equiv \frac{d}{dt}$, $\Psi_1 = H$, $f_1 \equiv 0$.

The system is Lagrangian and the motion is determined by vanishing of the forms

$$\begin{cases} \frac{\partial L}{\partial r} - \frac{\partial}{\partial t}\left(\frac{dL}{\partial \dot{r}}\right) \\ \frac{\partial L}{\partial \theta} - \frac{\partial}{\partial t}\left(\frac{\partial L}{\partial \dot{\theta}}\right) \end{cases}$$

where $L = \frac{1}{2}(mr^2\,\dot{\theta})\cdot\dot{\theta} + \frac{1}{2}(m\dot{r})\cdot\dot{r} - H.$

$$= \tfrac{1}{2}\{m(r\dot{\theta})^2 + m(\dot{r})^2\} - \tfrac{1}{2}Kr^2 - mg(h_0 - r\sin\theta).$$

The Hilbert spaces H_1 and H_2 are the trivial spaces $H_1 = \mathbb{R}$ and $H_2 = \mathbb{R}$, since ψ_1 is a functional, and in this case A represents classical differentiation. A different set of physical laws may be postulated if external forces Q act on the pendulum. Let the $p_\theta = mr^2\dot{\theta}$, $p_r = m\dot{r}$ be the angular and linear momentum. Let the physical laws stated for this problem be the Newton's laws of motion. The rate of change of the generalized momentum is equal to the total generalized force applied, i.e.

$$\frac{d}{dt}(p_\theta) = F_\theta, \quad \frac{d}{dt}(p_r) = F_r.$$

We choose $\Psi_1 = p_\theta, \Psi_2 = p_r$, $A = \frac{d}{dt}$, $F_r = Kr + mr\cdot\dot{\theta}^2 - mg\sin\theta + Q_r; F_\theta = mg\,r\cos\theta + Q_\theta$.

That is, we simply add all forces in the r-direction - the force of the linear spring Kr, the centrifical force $mr\dot{\theta}^2$ the force due to gravity - mg sin θ, and similarly add the two forces acting in the θ direction. We assume Sobolëv structure H^1 for the momenta p_θ, p_r, and for the displacements $\underset{\sim}{q} = \begin{bmatrix} q_r \\ q_\theta \end{bmatrix}$. The operator A maps H^1 into itself. The system is not Lagrangian, unless the forces Q_r, Q_θ may be derived from a potential U, i.e. if

$Q_r = \frac{\partial U}{\partial r}$, $Q_\theta = \frac{\partial U}{\partial \theta}$. In that case the previously defined Lagrangian L has to be replaced by $\mathcal{T}-U = L_1$.

Then Newton's laws can be derived by equating to zero the Euler-Weierstrass forms E_r, E_θ. The performance of the systems given in examples (2a), (2b),...etc could be the **real maximum energy** level attained by the system on some given time interval, the minimum displacement, the minimum of Kinetic energy, or almost any functional assigned to the system which measures some important physical property of that system.

<u>Lagrangian systems</u>

<u>Definition.</u> A system is called Lagrangian if for every admissible motion q(t) there exists a map $\Psi_t(q,\dot{q})$ called the Lagrangian, such that for any $t_1 \in [0, \infty]$ and $\eta \in \mathcal{M}$, (while Ψ is Fréchet differentiable at t_1) the derivative

$$\Psi_q = \frac{\partial \Psi}{\partial \hat{q}} - \frac{d}{dt}\left(\frac{\partial \Psi}{\partial \dot{q}}\right)$$ vanishes along the admissible motion $q(t) = \hat{q}(q_0,t)$ with $\hat{q}(q_0,0) = q_0$, and Ψ obeys the state equation: $A\Psi = 0 \in TM^*$.

If $\Psi_{\underset{\sim}{q}} = \underset{\sim}{Q} \neq 0$ then the system is said to have non-Lagrangian forces. We observe that the Fréchet derivative of Ψ is derived from the bilinear form where $<\frac{\partial\Psi}{\partial\hat{q}}, \varepsilon\, \eta>_{H_1} + <\frac{\partial\Psi}{\partial\dot{q}}, \varepsilon\, \dot{\eta}>_{H_1} + \mathcal{O}(\varepsilon)$,

$\varepsilon\, \eta \in \mathcal{M}$
$\varepsilon\, \dot{\eta} \in \mathcal{M}$ for small values of ε.

We denote by E_k the Weierstrass-Euler form

$$E_k(\Psi(q,\dot{q})) = \frac{\partial\Psi}{\partial q_k} - \frac{\partial}{\partial t}\left(\frac{\partial\Psi}{\partial\dot{q}_k}\right) \in TM^* .$$

The mapping $A : H_1 \to H_2$ is called non-Lagrangian at $t=t_1$ if $AE_k(\Psi)_{t_1} \neq 0$, for some k, $1 \leq k \leq n$. We assume that $\frac{\partial\Psi}{\partial q_k} \in H_1$, that $\frac{\partial}{\partial t}\left(\frac{\partial\Psi}{\partial\dot{q}_k}\right) \in H_1$, that is, we assign a Hilbert space structure to some elements of TM^*, while A maps H_1 into H_2. Specifically, we stipulate that the image of the Weierstrass-Euler form E_k is not the zero vector in H_2.

Let A be a linear map from H_1 to H_2 defined on H_1. Let A^* be the ajoint map-A^*: $H_2 \to H_1$.

Theorem. (Vaînberg) Let $\Psi(q,\dot{q})$, $\Phi(q)$ be functionals defined on $\mathcal{M}$, such that

$\frac{\partial\Psi}{\partial q}$, $\frac{\partial\Phi}{\partial q}$ ε TM^* are defined for all admissible motions $q(q_0,t)$, $t \in [0,T]$. A necessary condition for the locally extremal property of $\Phi(\tilde{q})$, where $\tilde{q}$ obeys the state equation $A\,\psi(q) = 0$, and A is a smooth linear operator A: $H_1 \to H_2$,

is: $$\frac{\partial}{\partial \underset{\sim}{q}} (\Phi + < A\Psi(\underset{\sim}{q}), \underset{\sim}{\lambda} >) = 0,$$

$$\frac{\partial}{\partial \underset{\sim}{\lambda}} (\phi + < A\Psi(\underset{\sim}{q}), \underset{\sim}{\lambda} >) = 0,$$

where $\lambda \in TM$ is an element of H_2.

Lemma 1. Let us consider a family of motions of a mechanical system defined on a Kinematic manifold $\mathcal{M}$ is some neighborhood of a point $\underset{\sim}{q}_0 \in \mathcal{M}$. We assume that at least one of the Euler-Weierstrass forms $E_k(\psi(\underset{\sim}{q},\underset{\sim}{q}))$ does not vanish in some neighborhood $N(q_o)$ of $\underset{\sim}{q}_0 \in \mathcal{M}$, where $\Psi(\underset{\sim}{q},\dot{\underset{\sim}{q}})$ is a smooth function of $\underset{\sim}{q}$ and $\dot{\underset{\sim}{q}}$ which is an element of a Hilbert space H_1. Let Φ be a differentiable function of $\underset{\sim}{q} \in \mathcal{M}$ which is an element of H_1. Let A be a linear operator H_1 to a Hilbert space H_2, the domain of A dense in H_1. Let us assume that $E_k^{-1} \frac{\partial \Phi}{\partial q_k}$ is in the domain of A. Let A* be the adjoint of A whose range is all of H_1, the domain all of H_2. Then there exists $\lambda_k \in H_2$(in the domain of A*) such that$\{E_k(\Psi(\underset{\sim}{q},\dot{\underset{\sim}{q}}))$. $A^*\lambda_k\} =$

$$\frac{\partial \Phi}{\partial q_k} = 0 \quad \text{in } N(\underset{\sim}{q}_0).$$

Proof: We denote by B the operator AA*. By our assumptions B^{-1} exists. Let us introduce

$\lambda_k \in H_2$, such that

$$\lambda_k = B^{-1}(E_k^{-1}(\psi) \frac{\partial \Phi}{\partial q_k}) \text{ in } N(q_0)$$

Then $A^* \lambda_k = A^* B^{-1} A(E_k^{-1}(\Psi) \frac{\partial\Phi}{\partial q_k})$

and

$$AA^* \lambda_k = AA^* B^{-1} A(E_k^{-1}(\Psi) \frac{\partial\Phi}{\partial q_k})$$

$$= A(E^{-1}(\Psi) \frac{\partial\Phi}{\partial q_k}),$$

i.e.

$$A(A^* \lambda_k - E_k^{-1} (\Psi) \frac{\partial\Phi}{\partial q_k}) = 0 \in H_2.$$

Since range of A^* is all of H_1, the null space of A consists of the zero vector only. Hence, the following equality is valid

$$A^* \lambda_k = E_k^{-1} (\Psi) \frac{\partial\Phi}{\partial q_k} \in H_1$$

and we have the equality

$$E_k(\Psi(\underset{\sim}{q},\underset{\sim}{\dot{q}}) A^* \lambda_k = \frac{\partial\Phi}{\partial q_k},$$

valid in $N(\underset{\sim}{q}_0) \in TM^*$,

which was to be proved.

The Concept of Sensitivity

Let us assume that the functions (or functionals) $\Psi_\alpha(\underset{\sim}{\tilde{q}})$, $\alpha = 1,2..k$, are defined on the Kinematically admissible manifold $\mathbf{M}$ of a mechanical system, $\Psi_\alpha: \mathbf{M} \to \mathbf{R}$. Each Ψ_α is also an element of a Hilbert space H_1, with the inner product

$\langle,\rangle_{H_1}$. A: $H_1 \to H_2$ is a smooth linear mapping, which depends on the design vector $\underset{\sim}{s} \in S \subset \mathbf{R}^m$. The derivative A_s of the operator A with respect to $\underset{\sim}{s}$ exists in some neighborhood of a design

$\underset{\sim}{s}_0 \in S$. For a group of local transformations $\Gamma: S \to S$(near s_0) A_s is the infinitesimal generator of Γ if we identify s_0 with the origin.

We assume the invariance of the physical laws: $A\,\Psi_\alpha(\underset{\sim}{q},\dot{\underset{\sim}{q}},s) = f_a(\underset{\sim}{q},\dot{\underset{\sim}{q}})$, $\alpha = 1,2,\ldots k$, under the action of the group. We shall treat two separate cases.

a) Lagrangian systems.

Let $\Psi_1(\underset{\sim}{q},\dot{\underset{\sim}{q}})$ denote the Lagrangian. Hence, for any $\underset{\sim}{s} \in S$, the motion of the system is determined by the vanishing of the Euler-Weierstrass forms $E_k(\Psi_1(\underset{\sim}{q},\dot{\underset{\sim}{q}})) = 0$, $k = 1,2,\ldots n$, in the local coordinate atlas $\{\underset{\sim}{q}\}$ covering $\mathbf{M}$. The invariance of the physical law $A\Psi_1 = 0$ is expressed by

$$\delta \underset{\sim}{s}_0 \int^T \left[\frac{d}{d\underset{\sim}{s}}\left(\frac{\partial \Psi_1}{\partial \underset{\sim}{q}} - \frac{\partial}{\partial t}\frac{\partial \Psi_1}{\partial \dot{\underset{\sim}{q}}}\right)\right] dt = 0, \text{ or}$$

$$\delta \underset{\sim}{s}_0 \int^T \left[\left(\frac{d}{d\underset{\sim}{s}} A\right) \Psi_1 + A\frac{d\Psi_1}{d\underset{\sim}{s}}\right] dt =$$

$$\delta \underset{\sim}{s}_0 \int^T \left[\left[A_{\underset{\sim}{s}}\, \Psi_1 + A\frac{\partial \Psi_1}{\partial \underset{\sim}{s}} + \frac{\partial \Psi_1}{\partial \underset{\sim}{q}}\frac{\partial \phi(\underset{\sim}{s},t)}{\partial \underset{\sim}{s}} + \right.\right.$$

$$\left.\frac{\partial \Psi_1}{\partial \dot{\underset{\sim}{q}}}\frac{\partial \dot{\phi}(s,t)}{\partial \underset{\sim}{s}}\right] dt = \left[\frac{d}{d\underset{\sim}{s}} \int_0^T (A\Psi_1)dt\right]\delta \underset{\sim}{s} = 0,$$

where $\underset{\sim}{\phi}(\underset{\sim}{s},t) = \underset{\sim}{q}(\underset{\sim}{q}_0(s), \underset{\sim}{s}, t)$ is the trajectory of the system obeying the initial conditions $q(\underset{\sim}{s},0) = q_0$, that are independent of $\underset{\sim}{s}$, and

$\frac{\partial \phi}{\partial \underset{\sim}{s}}$ is the matrix of partial derivatives

$$\left[\frac{\partial \phi_i}{\partial s_j}\right], \quad \left.\begin{array}{l} i = 1,2,\ldots n, \\ j = 1,2,\ldots k. \end{array}\right\}$$

A is a linear operator that maps H_1 into a Hilbert space H_2. Φ is a quality functional (or cost functional) associating with each motion $\{\hat{\phi}(t) = \phi(\underset{\sim}{s},t): \mathbf{R} \to \mathbf{M}\}$ a real number $\Phi(\hat{\Phi}(t))$.

The problem of improving an engineering design consists of estimating the sensitivity of Φ with respect to design changes in the space S of admissible designs.

Let us assume that the function $\underset{\sim}{\Psi}$ is uniformly non-Lagrangian on $\mathbf{M}$, i.e. along the entire trajectory

$$\phi(\underset{\sim}{s},t) = \hat{\phi}(t): \mathbf{R}_+ \to \mathbf{M}, \text{ such that } \phi(\underset{\sim}{s},0) = q_0,$$

the Euler-Weierstrass forms $E_K(\underset{\sim}{\Psi}(\underset{\sim}{q},\dot{\underset{\sim}{q}}))$ are all of constant sign. We also need the following properties of the operators A and A*. A* exists, is defined on all of H_2 and its range is all of H_1. That is, we assume that the hypothesis of lemma 1 are satisfied. Consequently, an element $\underset{\sim}{\lambda} \in H_2$ exists such that

$$A^* \lambda_K = E^{-1}(\underset{\sim}{\Psi})\frac{\partial \Phi}{\partial q_k}, \quad k = 1,2,..n.$$

We wish to derive a sensitivity formula for $\frac{d\Phi}{d\underset{\sim}{s}}$. Again, we postulate invariance of physical laws, or constitutive relations.

Hence, for the action integral the sensitivity is given by

$$(3.2)\quad \frac{d}{d\underset{\sim}{s}}\int_0^T (A\Psi_1)\,dt = \int_0^T [(A_s\,\Psi_1) + (A\frac{\partial\Psi_1}{\partial \underset{\sim}{s}})]\,dt$$

$$+ \int_0^T \{[\frac{\partial\Psi_1}{\partial \underset{\sim}{q}} - \frac{\partial}{\partial t}(\frac{\partial\Psi_1}{\partial\dot{q}})]\frac{\partial\underset{\sim}{\phi}}{\partial \underset{\sim}{s}}\}\,dt\,.$$

$$+ [\frac{\alpha\Psi_1}{\alpha\dot{\underset{\sim}{q}}}\cdot\frac{\partial\phi}{\partial \underset{\sim}{s}}] = \int_0^T\left[(A_{\underset{\sim}{s}}\,\Psi_1) + A(\frac{\partial\Psi_1}{\partial \underset{\sim}{s}})\right]dt +$$

$$\left[\frac{\partial\Psi_1}{\partial\dot{\underset{\sim}{q}}}\,\frac{\partial\phi}{\partial \underset{\sim}{s}}\right]_{t=0}^{t=T}\,.$$

We observe that the second integral in (3.2)

vanishes because $\frac{\partial\Psi_1}{\partial \underset{\sim}{q}} = \frac{\partial}{\partial t}(\frac{\partial\Psi}{\partial\dot{\underset{\sim}{q}}})$

along a trajectory $\underset{\sim}{q}(t) = \phi(\underset{\sim}{s},t) \in \mathcal{M}$.

This relation allows us to compute the sensitivity of the solution $\phi(\underset{\sim}{s},t)$ at the final instant $t = T$. Specifically, if $\phi_0 = \phi(\underset{\sim}{s},0)$ is given (independently of $\underset{\sim}{s}$) then

$$(3.3)\quad \frac{d}{d\underset{\sim}{s}}\phi(\underset{\sim}{s},t) = -\int_0^T \left\{ (A_{\underset{\sim}{s}}\,\Psi_1) + A\frac{\partial\Psi_1}{\partial \underset{\sim}{s}} \right\} dt,$$

plus a term that depends only on the final condition (at $t = T$) of the system.

In some problems the optimization of that term is most significant.

The importance of this result is best illustrated with a simple physical example.

Consider a (conservative) connected pendulum system performing small vibrations about the origin $\{\theta_1=0,\theta_2 = 0\}$.

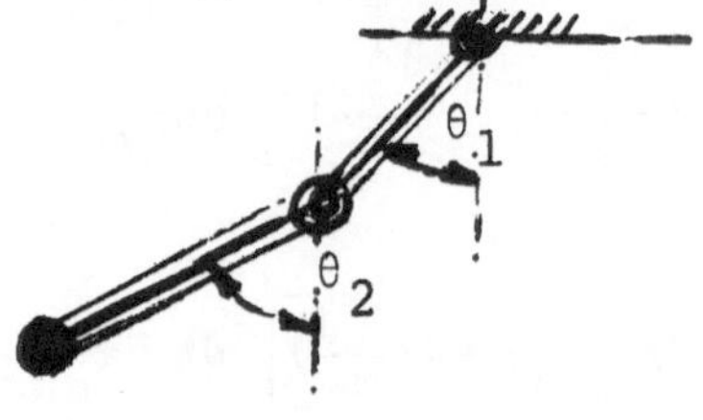

Figure 3.5

The Lagrangian is given by:

$$(3.4)\quad L(\theta_1,\theta_2) = \tfrac{1}{2}\,\ell^2(m_1\dot{\theta}_1{}^2+m_2\dot{\theta}_2{}^2) - g\ell(m_1(1-\cos\theta_1) + m_2(1-\cos\theta_2))+\tfrac{1}{2}\,\ell^2(\theta_1-\theta_2)^2$$

L is a function of $\{\theta_1,\dot{\theta}_1,\ \theta_2,\ \dot{\theta}_2\}$ and of the design parameters m_1, m_2, ℓ and K.

The design vector is $s = \{m_1, m_2, \ell, K\}$.

The motion of the system in the absence of external forces is given by Newton's second law

$A(\mathbf{L}(\theta_1,\theta_2)) =$

$$(3.5)\quad \left.\begin{aligned} &\frac{\partial L}{\partial \theta_1} - \frac{d}{dt}\left(\frac{\partial L}{\partial \dot{\theta}_1}\right) = 0, \\ &\frac{\partial L}{\partial \theta_2} - \frac{d}{dt}\left(\frac{\partial L}{\partial \dot{\theta}_2}\right) = 0. \end{aligned}\right\}$$

i.e.

$$\frac{d}{dt}\left(m_1 \ell^2 \frac{d\theta_1}{dt}\right) - g\ell\, m_1 \sin\theta_1$$

$$+ \ell^2(\theta_1 - \theta_2) .$$

$$A(\ell) = A(\theta(\ell), \theta_2(\ell), \ell) =$$

$$\frac{d}{dt}\left(m_2\, \ell^2 \frac{d\theta_2}{dt}\right) - g l m_2 \sin\theta_2$$

$$+ \ell^2 (\theta_2 - \theta_1),$$ where

the operator $A = \begin{cases} \frac{\partial}{\partial \theta_1} - \frac{\partial}{\partial t}\left(\frac{\partial}{\partial \dot{\theta}_1}\right) \\ \frac{\partial}{\partial \theta_2} - \frac{\partial}{\partial t}\left(\frac{\partial}{\partial \dot{\theta}_2}\right) \end{cases}$

does not explicitely depend on $\underset{\sim}{s}$. Without computing a solution to the system of differential equations (3.5) we wish to find the sensitivity of that solution at time t = T for a given design $\underset{\sim}{S} = \underset{\sim}{S}_0$.

The initial conditions assigned to this problem are independent of s.

We have $\frac{d}{ds}\left[\hat{\theta}_1(\underset{\sim}{s},t), \hat{\theta}_2(\underset{\sim}{S},t)\right]_{t=T} =$

(3.6) $$\int_0^T \left(A \frac{\partial L}{\partial \underset{\sim}{s}}\right) dt = \begin{bmatrix} \int_0^T \left(A \frac{\partial L}{\partial m_1}\right) dt \\ \int_0^T \left(A \frac{\partial L}{\partial m_2}\right) dt \\ \int_0^T \left(A \frac{\partial L}{\partial \ell}\right) dt \\ \int_0^T \left(A\left(\frac{\partial L}{\partial K}\right)\right) dt \end{bmatrix}$$

Hence, $\frac{d}{dm_1}[\hat{\theta}_1,\hat{\theta}_2]_{t=T} = \int_0^T(\ell^2\ \ddot{\hat{\theta}}_1 - g\ell\ \sin\theta_1)\ dt$

For small oscillations about the origin this integral is approximated by

$$\int_0^T(\ell^2\ \ddot{\hat{\theta}}_1 - \ell\ g\hat{\theta}_1)\ dt,$$

computed for the design $\underset{\sim}{s} = \underset{\sim}{s}_o$, that is for $\ell = \ell_0$.

Similarly we compute the sensitivity with respect to m_2, m_1,K. Hence, it suffices to produce numerically a solution $\{\hat{\theta}_1, \hat{\theta}_2\} = \hat{\underset{\sim}{\theta}}$ for a fixed (given) design $s \in S$ and to evaluate a rather simple integral along that trajectory, with $\hat{\underset{\sim}{\theta}}$ satisfying initial conditions at t = 0, in order to determine the sensitivity of the solution at a given time t = T. Solution of the adjoint system was not needed. We observe that group theoretic arguments were not required and the topological properties of the Kinematic manifold were used without explicitely defining its basic topological structure.

That is, in the numerical solution $\theta_1 = \theta_1 \pm 2\pi n$, $\theta_2 = \theta_2 \pm 2\pi n$. Elastic coupling would not influence the admissible configuration of the Kinematic manifold, which is a torus. That is replacing a bearing in suspension of the second (lower) pendulum in figure (3.5) by a torsion spring does not change the topology of the kinematic manifold, but replacing the pendulum rods by elastic rods, that is by springs does. Instead of the Cartesian product of two circles (i.e. a torus) we obtain a Cartesian product of two (2-dimensional) cylinders in $\mathbb{R}^4$.

3b. Non - Lagrangian systems.

Here we consider a mechanical system in a coordinate atlas q covering the admissible Kinematic manifold $\mathcal{M}$. A function Ψ: $\mathcal{M} \times TM \to \mathbb{R}^{\ell}$ and a functional Φ: $\mathcal{M} \to \mathbb{R}$ are given. The physical laws or constitutive equations are given by the operator equation $A\underset{\sim}{\Psi} = 0$ (or $A\underset{\sim}{\Psi} = \underset{\sim}{f}$), where is regarded as an element if a Hilbert space H_1,

$$(3.7)\quad \frac{d(A\Psi)}{d\underset{\sim}{s}} = A_{\underset{\sim}{s}}\Psi + A\{\frac{\partial\Psi}{\partial\underset{\sim}{s}} + \frac{\partial\Psi}{\partial\underset{\sim}{q}}\frac{\partial\Phi}{\partial\underset{\sim}{s}} + \frac{\partial\Psi}{\partial\dot{\underset{\sim}{q}}}\frac{\partial\dot{\Phi}}{\partial\underset{\sim}{s}}\} = 0 .$$

We invoke lemma 1.

An adjoint variable A^* is introduced which obeys the vector equation

$$(3.8)\quad A^* \underset{\sim}{\lambda} \left(\frac{\partial\Psi}{\partial\underset{\sim}{q}} - \frac{\partial}{\partial t}\frac{\partial\Psi}{\partial\dot{\underset{\sim}{q}}}\right) = \frac{\partial\Phi}{\partial\underset{\sim}{q}} .$$

$A^* : H_2 \to H_1$ is the adjoint of A. We have

$$\frac{d\Phi}{d\underset{\sim}{s}} = \frac{\partial\Phi}{\partial\underset{\sim}{q}}\frac{\partial q}{\partial\underset{\sim}{s}} + \frac{\partial\Phi}{\partial\underset{\sim}{s}} .$$

Hence,

$$(3.9^a)\quad \frac{\partial\Phi}{\partial\underset{\sim}{q}}\ \frac{\partial\Phi}{\partial\underset{\sim}{s}} = \frac{d\Phi}{d\underset{\sim}{s}} - \frac{\partial\Phi}{\partial\underset{\sim}{s}} .$$

The adjoint variable λ is introduced into the weak form of (3.5)

$$0 = \langle A_{\underset{\sim}{s}}^{*}\ \underset{\sim}{\lambda}, \underset{\sim}{\Psi}\rangle_{\Omega} + \langle A^*\ \underset{\sim}{\lambda}, \frac{\partial\Psi}{\partial\underset{\sim}{s}}\rangle_{\Omega}$$

$$+ \langle A^*\underset{\sim}{\lambda}\frac{\partial\Psi}{\partial q}, \frac{\partial\Phi}{\partial\underset{\sim}{s}}\rangle_{\Omega} - \langle A^*\underset{\sim}{\lambda}\frac{\partial}{\partial t}\left(\frac{\partial\Psi}{\partial q}\right), \frac{\partial\Phi}{\partial\underset{\sim}{s}}\rangle_{\Omega}$$

If the functional Φ is only implicitely dependent on the design $\underset{\sim}{s}$ we have

$\frac{\partial \Phi}{\partial \underset{\sim}{s}} = 0$ and

$$(3.9^c) \quad \frac{d\Phi}{d\underset{\sim}{s}} = < A_{\underset{\sim}{s}}^{*} \lambda, \Psi> + < A^{*}\lambda, \frac{\partial \Psi}{\partial \underset{\sim}{s}} >$$

If also Ψ $(q, \dot{q}, \underset{\sim}{s})$ does not explicitely depend on $\underset{\sim}{s}$, i.e. $\Psi = \Psi(q, \dot{q})$,

then

$$(3.10) \quad \frac{d\Phi}{d\underset{\sim}{s}} = <A_{\underset{\sim}{s}}^{*} \; \underset{\sim}{\lambda}, \underset{\sim}{\Psi} \; >.$$

Many problems of classical and continuum mechanics are reducible to the form $AA^{*}\omega = q$ or $AA^{*}\omega + \lambda \, BB^{*}\omega = q$. (3.11).Examples of such problems and related variational problems were discussed in chapters 1-4 of Volume 1.

In general, both the operators A, A* and the state function ω depend on a design vector u.

The general problems of continuity and differentiability are difficult to classify. However, for the operators of the type (3.11) encountered in continuum mechanics several important results are available in the literature. The eigenvalue problems present greater difficulties than the purely static problems in which the primary consideration is differentiability of the state operator with respect to the design vector u. Even then several distinct cases must be considered. In cases of finite element approximations, or of lumped parameters u may be regarded as a vector in $\mathbb{R}^k$, where k is geñerally quite large, but still, one deals with a finite-dimensional vector space U of admissible designs. If the problem is self-adjoint (A = A*), then the differentiability of a simple eigenvalue is easily established and it is given by

$$\frac{\partial \lambda}{\partial \underset{\sim}{u}} = < \frac{\partial \, A(\underset{\sim}{u})}{\partial \, \underset{\sim}{u}} \; x_i(u), \; x_i(u)>$$

for the eigenvalue problem (3.11): $Ax_i = \lambda_i x_i$.

Observe that

$$\lambda_i(u) = \langle A(u)\, x_i(u),\, x_i(u) \rangle / \langle x_i(u),\, x_i(u) \rangle$$

where x_i is corresponding eigenfunction inserted into the Rayleigh quotient.

We can assume that the quadratic form $\langle x_i, x_i \rangle \equiv 1$ (independently of $\underset{\sim}{u}$), so that (3.11) is rewritten as

$$\lambda_i = \min_{\phi(x) \in H} \langle A\, \phi, \phi \rangle = \langle A\, x_i, x_i \rangle \qquad (3.12)$$

We formally use the chain rule formula

$$\langle \frac{\partial A}{\partial u} x_i, x_i \rangle + \langle A \frac{\partial x}{\partial u}, x_i \rangle + \langle A\, x_i, \frac{\partial x_i}{\partial u} \rangle =$$

$$\frac{\partial x_i}{\partial u} \langle x_i, x_i \rangle + \lambda \langle \frac{\partial x_i}{\partial u}, x_i \rangle + \lambda \langle x_i, \frac{\partial x_i}{\partial u} \rangle$$

Thus, we have

$$\frac{\partial x_i}{\partial \underset{\sim}{u}} = \langle \frac{\partial A}{\partial \underset{\sim}{u}} x_i, x_i \rangle \quad (3.13).$$

In manipulating the equality

$$\langle A \frac{\partial x_i}{\partial \underset{\sim}{u}}, x_i \rangle = \lambda_i \langle \frac{\partial x_i}{\partial \underset{\sim}{u}}, x_i \rangle$$

we have used the self adjoint property of A and the fact that the eigenvalue λ_i is real.

The formula (3.13) is essential in computing sensitivity of a simple eigenvalue in many problems of engineering mechanics. The development given above was purely formal. We did not worry about the properties of the spaces containing eigenfunctions x_i or approximate eigenfunctions and functions Ax_i as vectors, nor was the symbol $\frac{\partial A}{\partial \underset{\sim}{u}}$ properly defined.

Neither the domain, codomain or range of $\frac{\partial A}{\partial \underset{\sim}{u}}$ was discussed, nor was its existence proved.

Of course, in general, it is not possible to prove its existence and specific cases must be considered in detail. The derivative $\frac{\partial A}{\partial \underset{\sim}{u}}$ of the operator A with respect to $\underset{\sim}{u}$ is not just another operator mapping $H_1 \rightarrow H_2$. In the simplest case when $\underset{\sim}{u}$ is a finite-dimensional vector of scalars ($\underset{\sim}{u} \in R^k$, the operator $\left.\frac{\partial A}{\partial \underset{\sim}{u}}\right|_{\underset{\sim}{u}_o}$ is k-tuple of operators $H_1 \rightarrow H_2$, provided that each of these operators is defined. It is easy to exhibit an operator $\frac{\partial A}{\partial u_i}$ that is not defined for some value of the component u_i of $\underset{\sim}{u}$. For example, the operator: $A(a) = \frac{\partial}{\partial x} (\frac{1}{a^2}) \frac{\partial}{\partial x}$ depending on a single parameter a obviously does not have a derivative $\frac{\partial}{\partial a}$ in the neighborhood of $a = 0$. However, with suitable limitations placed on the values of the parameters it is possible to prevent such difficulties.

In fact, with some fairly obvious limitations, this formal approach can be extended to non-selfadjoint eigenvalue problems.

Basically, no serious problems arise if one assumes differentiability and regular behaviour of all terms. In considering the dependence of the eigenvalue on a design vector $\underset{\sim}{u}$, we can use the energy bilinear form equation

$$(3.14) \qquad \frac{\partial \lambda_i}{\partial \underset{\sim}{u}} = < \frac{\partial A}{\partial \underset{\sim}{u}} x_i, x_i > - \lambda_i < \frac{\partial B}{\partial \underset{\sim}{u}} x_i, x_i >$$

where as before λ_i is the eigenvalue, x_i the

corresponding eigenfunction obeying

(3.15) $Ax_i = \lambda Bx_i$;

with x_i normalized :

$< Bx_i, x_i > = 1.$

Here A, B are selfadjoint, operators, and λ_i a simple eigenvalue. Let us offer some simple examples of application before we discuss the difficulties and reasons why this formula may be incorrect.

a) Vibration of a beam built in at both ends results in the eigenvalue problem

$$\left.\begin{array}{l} L(w) = (EI(x)\, w_{xx})_{xx} = \lambda \rho\ A(x)\, w(x) \\ w(0) = w(\ell) = 0 \\ w_x(0) = w_x(\ell) = 0 \\ \text{where } \lambda = -\omega^2 \end{array}\right\} \quad (3.16)$$

where ρ is the density, assumed to be constant.

Let $I(x)$, $A(x)$ depend on some scalar function parameter $u(x)$, $u \in L^\infty[0,\ell]$, i.e. $I(x) = I(u(x))$, $A = A(u(x))$. It is impossible to deal directly with the operator L in (3.16)

It is unbounded and the chain rule cannot be applied without great misgivings.

However the corresponding bilinear energy forms

$$a(w,v) = \int_0^\ell (E\, I(x)\, w_{xx} \cdot v_{xx})\, dx = < EI\, w_{xx}, v_{xx} >$$

$$b(w,v) = \int_0^\ell (\rho\, A(x)\, w(x) \cdot v(x))\, dx = < \rho \cdot A\, w, v >,$$

are Fréchet differentiable under very mild

assumptions (see Kato [27], Haug [28], Haug and Rousselet [29]) for sufficiently smooth choices of test function v(x). The test function v(x) is chosen independently of the design $\underset{\sim}{u}(x)$.

Thus $\partial a/\partial u = E \{<\frac{\partial I}{\partial u} \omega_{xx}, v_{xx}> + < I(\frac{\partial \omega}{\partial u})_{xx}, v_{xx}>,$

$\partial b/\partial u = \rho \{<\frac{\partial A}{\partial u} \omega, v> + < A \cdot \frac{\partial \omega}{\partial u} v > \} .$

The chain rule for Fréchet differentiation gives us

$$(3.17)\quad \partial a/\partial u = \partial\lambda/\partial u \cdot b + \lambda\cdot\frac{\partial b}{\partial u}$$

or

$$(3.18)\quad < E \frac{\partial I}{\partial u} \omega_{xx}, v_{xx}> + < EI(\frac{\partial \omega}{\partial u})_{xx}, v_{xx}> =$$

$$\frac{\partial \lambda}{\partial u} < \rho A \omega, v> + \lambda \{<\rho \frac{\partial A}{\partial u} \omega, v> + <\rho A\frac{\partial \omega}{\partial u}, v>.$$

If $\omega(x)$ is sufficiently smooth, then there is no reason why it or $\frac{\partial \omega}{\partial u}$, could not be regarded as the test functions.

Let us identify ω with v, $\frac{\partial \omega}{\partial u}$ with a test function ϕ, and observe that

$$< EI(\frac{\partial \omega}{\partial u})_{xx}, v_{xx} > - \lambda<\rho A\frac{\partial \omega}{\partial u}, v > = < EI \phi_{xx}, \omega_{xx}>-$$

$$\lambda<\rho A\phi, \omega > = < EI \omega_{xx}, \phi_{xx} > - \lambda<\rho A\omega, \phi > = 0 , \quad (3.19)$$

if ω is a weak solution of the original eigenvalue problem. Thus, we rederive again the formula

$$<E \frac{\partial I}{\partial u} \tilde{\omega}_{xx}, \tilde{\omega}_{xx}> - \tilde{\lambda} < \rho \frac{\partial A}{\partial u} \tilde{\omega}, \tilde{\omega} > = \frac{\partial \tilde{\lambda}}{\partial u} <A\rho\tilde{\omega},\tilde{\omega} >$$

where $\tilde{\lambda}$ is an eigenvalue $(\tilde{\lambda} = -\omega^2)$ and $\tilde{\omega}$ corresponding eigenfunction of (3.16).

ω denotes the natural frequency.

(b) Vibration of a thin plate.

The energy bilinear form for the Lagrange-Germaine plate model is given by

$$a(u,v) = \iint_{\Omega} \{\frac{Eh^3}{12(1-\nu^2)} [(\Delta u \cdot \Delta v) - (1-\nu) \lozenge^4(u,v)\} \, dx\, dy, \qquad (3.20)$$

where $\Delta = \frac{\partial^2}{\partial x^2} + \frac{\partial^2}{\partial y^2}$, and $\lozenge^4(u,v) = \frac{\partial^2}{\partial x^2}\frac{\partial^2 v}{\partial y^2} - 2\frac{\partial^2 u}{\partial x \partial y}\frac{\partial^2 v}{\partial x \partial y} + \frac{\partial^2 u}{\partial y^2}\frac{\partial^2 v}{\partial x^2}$. As before, E is the Young modulus (a positive number) ν-is the Poisson ratio ($0 \leq \nu \leq \frac{1}{2}$); $h(x,y)$ is the plate's thickness; h is piecewise continuous, but we can assume that $h \in L_2(\Omega)$, $h_1 \leq h \leq h_2$, where h_1, h_2 are positive constants. The second fundamental energy form is

$$b(u,v) = \iint_{\Omega} (\rho \cdot h \cdot u \cdot v) \, dx\, dy, \qquad (3.21)$$

where ρ is the material density ($\rho > 0$).

Here we assume that ρ is constant, $h = h(x,y)$.

The weak formulation of the sensitivity for the eigenvalue λ of $a(u,v) = \lambda b(u,v)$ is

$$\frac{\partial\, a(u,v)}{\partial h} = \frac{\partial \lambda}{\partial h} b(u,v) + \lambda \frac{\partial\, b(u,v)}{\partial h}. \qquad (3.22)$$

$\frac{\partial \lambda}{\partial h}$ now assumes the form

$$\frac{\partial \lambda}{\partial h}(x,y) = \frac{Eh^2}{6(1-\nu^2)} [\Delta u \cdot \Delta v - (1-\nu) \cdot \lozenge^4(u,v)] + \lambda \cdot (\rho \cdot u \cdot v),$$

defining the (local) sensitivity of

the eigenvalue $\lambda = -\omega^2$ with respect to the single design parameter $h(x,y)$ identified with the thickness of the plate.

c) <u>A one-dimensional shape sensitivity problem.</u>

The vibrating string problem is given by the eigenvalue system

$$(3.23) \qquad y''(x) = (-\omega^2)\cdot y(x) \qquad x \in (0,\ell),$$

$$(y \not\equiv 0)$$

$$y(0) = 0,$$

$$y(\ell) = 0 .$$

The domain problem is the problem of determining sensitivity of the frequency with respect to ℓ. The equation (3.23) has the obvious solution

$$y = C \sin \omega x \qquad \text{with} \qquad \omega = \frac{\Pi}{\ell} .$$

If we regard ℓ as a design parameter, the sensitivity of the state is

$$\frac{dy(x)}{d\ell} = -\frac{2\Pi^2}{\ell^3} C \cos(\frac{\Pi}{\ell}x) = -\frac{2\Pi}{\ell^2}\left(\frac{dy(x)}{dx}\right) .$$

The sensitivity of the frequency is

$$\frac{d\omega}{d\ell} = -\frac{\Pi}{\ell^2}, \text{ and } \frac{d\lambda}{d\ell} = d\,\frac{(-\omega^2)}{d\ell} = \frac{-2\Pi^2}{\ell^3} .$$

Clearly, there was no need for the energy formula (3.22) to compute the sensitivity of the eigenvalue. However, we could duplicate the more difficult problems by rewriting this eigenvalue problem in the weak form, using an arbitrary test function $v \in H^1(0,\ell)$.

$$\int_0^\ell (w''v)\,dx = -\omega^2 \int_0^\ell (wv)\,dx ,$$

or

$$\int_0^\ell (w'v')\,dx = -\omega^2 \int_0^\ell (wv)\,dx = \lambda \int_0^\ell (wv)\,dx.$$

Therefore, $\dfrac{d\lambda}{d\ell} \displaystyle\int_0^\ell (wv)\ dx + \lambda\ \dfrac{d}{d\ell} \displaystyle\int_0^\ell (wv)\ dx$

$$= \dfrac{d}{d\ell} \int_0^\ell (w'v')\ dx,$$

Here we have the recall the rules of differentiation with respect to the limits of an integral. In the one-dimensional case the answer is fairly trivial.

$\dfrac{d}{db} \displaystyle\int_a^b y(x)\ dx = y(b)$, provided $y(x)$ is continuous at b. This is easily checked by considering the limit:

$$\lim_{\varepsilon \to 0_+} \dfrac{\displaystyle\int_a^{b+\varepsilon} y(x)\ dx - \int_a^b y(x)\ dx}{\varepsilon} = \lim_{\varepsilon \to 0} \dfrac{\displaystyle\int_b^{b+\varepsilon} y(x)\ dx}{\varepsilon}$$

$$= y(b).$$

Identifying v with w, and recalling that $w(\ell) = 0$, we obtain

$$\dfrac{\partial \lambda}{d\ell} = \dfrac{(w'(\ell))^2}{\displaystyle\int_0^\ell w^2 dx}\ .$$

This checks our previous formula if we substitute $w = \sin(\omega x)$, with $\omega = \dfrac{\pi}{\ell}$ and $\lambda = -\omega^2$. This was quite elementary. However, it is not as easy in the higher dimensional case, where perturbation of the domain has been regarded for some time as a difficult and subtle problem.

Higher dimensional problems.
Hadamard's approach

The free boundary problems are extensively discussed in Avner Friedmann's monograph [31].

The approaches taken by several authors including Friedmann use various energy criteria, symmetrization and functional analytic techniques to deduce some properties of the free boundary in Stefan's problem, or in related problems (such as contact elasticity) where the shape of the boundary is not known, but some energy criteria are available to compare the candidates for the actual (that is the "real", "physical") boundary shape. Armand attempted to convert the free boundary problem to an equivalent optimization problem, in which the change of the shape of the boundary becomes a variation of the coefficients in a partial differential equation. Most of the recent progress by the French school (See Cea [33], Pironneau [35], Cea, Gioan and Michel [36], Dervieux and Palmerio [37], Dervieux [38]) relied heavily on the idea of Hadamard first published in 1910 [39].

Let Γ_0 be a position of the boundary and Γ_ε a perturbed boundary given by

$\Gamma_\varepsilon = \{ x + \varepsilon \mathbf{n} \cdot \mu(x) \mid x \in \Gamma_0\}$. Let us suppose that we know the Green's function for the problems

$L(u(\underset{\sim}{x}) = f$, and it is given by $G(\underset{\sim}{\xi}, \underset{\sim}{\mathbf{n}}, \Gamma_0)$. Its perturbation is $\partial_\varepsilon G = \frac{d}{d\varepsilon} G(\underset{\sim}{\xi}, \underset{\sim}{\mathbf{n}}, \Gamma_\varepsilon)\big|_{\varepsilon=0}$.

This is given by

$$\int_\Gamma^0 - \frac{\partial G(\xi, s, \Gamma_0)}{\partial \mathbf{n}_s} \quad \frac{\partial G(s, \mathbf{n}, T_0) \cdot \mu(s)}{\partial \mathbf{n}_s} \, d\Gamma(s)$$

for the Dirichlet problem.

Chapter 3

References for Chapter 3

[1] I.M. Gelfand and G.E. Shilov, Volume I of Generalized Functions, Academic Press, New York, 1964. Translation from Russian by George Sal etain.

[2] A. Zemanian, Distribution theory and transform analysis, McGraw Hill, New York, 1965.

[3] L. Schwartz, Theory of distributions, Vol. I and II, Herman & Cie, Paris, 1957 and 1959.

[4] P.G. Drazin, Solitons, Cambridge University Press, London Mathematical Society Lecture Series #85, Cambridge University Press, 1983.

[5] R.M. Miura, C.S. Gardner and M.D. Kruskal, Kerteweg de Vries equation and generalizations, J. Math. Physics, #9, 1968, p. 1204-1209.

[6] R.M. Miura, The Korteweg-de Vries equation. A survey of results, S.I.A.M. Review, 18, (1976), p. 412-459.

[7] M. Toda, Wave propagation in anharmonic lattice, J. Phys. Soc. Japan, 23, (1967), p. 501-506.

[8] M.V. Novitskii, On reconstruction of the rotation number function for the Schrödinger operator with almost periodic potential on a denumerable set of polynomial conservation laws; Functional Analysis and its Applications, Vol. 19, (1985) #3, p. 90-91.

[9] M. Crampin, Constants of motion in Lagrangian mechanics, Int. J. theoret. physics, Vol. 16, #10 (1977) p. 741-754.

[10] V.I. Arnol'd, Mathematical methods of classical mechanics, Springer-Verlag, New York, Heidelberg, Berlin, 1978.

[11] M. Spivak, Differential geometry (five volumes) Publish or Perish Press, Boston, Mass, 1979.

[12] B.A. Dubrovin, A.T. Fomenko and S.P. Novikov, Modern geometry - methods and applications, Vol. I, Springer-Verlag, Graduate Texts in Mathematics, Vol. 93,1984.

[13] D. Lovelock and H. Rund, Tensors, differential forms and variational principles, Wiley-Interscience, New York, 1975.

[14] W. Thirring, A course in mathematical physics, Vol I, classical dynamical systems, Springer-Verlag, New York, 1978.

[15] R. Hermann, Vector bundles in mathematical physics, Vol. 1, Benjamin, New York, 1970.

[16] R. Hermann, Lectures in mathematical physics, Vol. I and II,

[17] N. Steenrod, Topology of fibre bundles, Princeton University Press, 1952.

[18] E. Cartan, Leçons sur la geometrie des espaces de Riemann, Gauthier-Villars, Paris, 1925.

[19] V.E. Stepanov and Iu. Ia. Iappa, Lie derivative of a geometric object, (towards a tetradic formulation of the theory of gravitation), Vestnik Leningrad. Univ., 22, (1978), p. 42-46.

[20] Felix Klein, Elementary mathematics from an advanced standpoint- Geometry, Dover Publ., New York, 1939(original publication, Berlin 1908).

[21] A. Lichnerowicz, Geometrie de Groupes des Transformations, Dunod, Paris, 1958.

[22] R. Abraham and J. Marsden, Classical mechanics, Springer New York, Berlin & Heidelberg, 1978.

[23] R. Saeks, Resolution spaces, Operators and systems, Lecture notes in Economics and Mathematical Systems, Vol. 82, Springer Verlag, Berlin, 1973.

[24] V. Komkov, Control theory, variational principles and optimal design of elastic systems, Proceedings of International Conference in Norman, Oklahoma, March 1977, J. Wiley & Son, New York, 1978.

[25] V. Komkov, Sensitivity analysis in some engineering applications, in Sensitivity of functionals with applications to engineering sciences, V. Komkov-editor, Lecture Notes in Mathematics Series,#1086, Springer Verlag, Berlin - Heidelberg - New York, 1984.

[26] E. Hille and R. Phillips, Functional Analysis and Semi-groups. Colloq. Publ. Amer. Math. Soc., 1957, Providence, R.I.

[27] T. Kato, Perturbation theory for linear operators, Springer Verlag, Berlin, 1967.

[28] E.J. Haug, A review of distributed parameter structural optimization literature, in volume 1 of Optimization of distributed parameter structures, NATO Advanced Study

[29] E. Haug and B. Rousselet, Design Sensitivity Analysis in Structural Mechanics, Part II, eigenvalue variations, J. Structural Mechanics, vol. 8, #2, 1980.

[30] H. Poincaré, Sur une forme nouvelle des equations de la mechanique, Comptes Rendues Acad. Sci. Paris, Vol. 132 (1901) p. 369-371.

[31] Avner Friedman, Variational Principles and Free Boundary Problems, Wiley-Interscience, New York, 1982.

[32] J.L. Armand, Numerical solutions in optimization of structural elements, TICOM First International conference on computing methods in nonlinear mechanics, J.T. Oden editor, Austin Texas, 1974.

[33] J. Cea, A numerical method for the computation of an optimal domain, Lecture Notes in Computer Science, Vol. 41, Springer Verlag, New York and Heidelberg, 1976.

[34] N. Banichuk, Problems and Methods of Optimal Structural Design, Plenum Press, New York, 1983.

[35] D. Pironneau, Optimal shape design for elliptic systems, Springer Verlag series in computational physics, New York, 1983.

[36] J. Cea, A. Gioan and J. Michel, Some results on domain identification problem, Calcolo Fasc. #3/4 (1973).

[37] A. Dervieux and B. Palmerio: A formula of Hadamard for domain identification problems (in French). Note CRAS A280(1975).

[38] A. Dervieux: A perturbation study of a jet-like annular free boundary problem. Comm. PDE 6(2)(1981).

[39] T. Kato, On some approximate methods concerning operators T*T Math. Ann., Vol. 126, (1955), p. 253-257.

Appendix A to Chapter 3

Lie groups and Lie algebras.

Elements of a Lie group **G** comprise a differentiable manifold. The group operation ∘ is a binary differentiable map of **G** × **G** into **G**.

An algebra on a vector space X with a multiplication operation $[\cdot,\cdot]: X \times X \to X$ which is linear and skew symmetric is called a Lie algebra if it satisfies the Jacobi identity

$$[\,[x,y],\, z] + [\,[y,z],\, x] + [\,[z,x],\, y] = 0.$$

The tangent space to a Lie group can be given a "natural" Lie algebra. A Lie algebra that is very important in mechanics is generated by the Poisson bracket that is commonly defined by a skew symmetric product

$$\{f,g\} = \sum_i \left(\frac{\partial f}{\partial x^i}\frac{\partial g}{\partial p_i} - \frac{\partial f}{\partial p_i}\frac{\partial g}{\partial x^i}\right)$$

The Poisson bracket satisfies the Jacobi identity thus generating a Lie algebra. We can also introduce the Lagrange brackets of parameters ξ,η which is:

$$\{\xi,\eta\}_{\underset{\sim}{x},\underset{\sim}{p}} = \sum_i \left(\frac{\partial x^i}{\partial \xi}\frac{\partial p_i}{\partial \eta} - \frac{\partial p_i}{\partial \xi}\frac{\partial x^i}{\partial \eta}\right).$$

where, clearly

$$\{\xi,\eta\}_{x,p} = -\{\eta,\xi\}_{x,p} .$$

This bracket will not be used presently, but it is of importance when effects of parameters are studied.

Appendix B

<u>A comment on some operators of mechanics.</u>

In many problems the operators governing the system are positive definite (or negative definite). For example the Laplace operator, the biharmonic operator, the Bernoulli beam operator, or the Sophie Germaine plate operator are all of the form A^*A. It is appropriate to summarize some of their properties.

Some easy to prove properties of the operator AA^* are listed below:

a) AA^* is obviously symmetric.

b) If A^* is a closed operator, then AA^* is a self-adjoint operator.

c) AA^* is non-negative

d) It is a positive definite operator if the null space of A^* consists of the zero vector only. (That is, for any $\xi \neq \emptyset \in D_{A^*}$ it is true that

$\langle AA^* \xi, \xi \rangle > 0$.) We assume that A^* is closed.

e) Every positive definite operator can be represented in the form AA^*.

f) Every operator of the form ABA^* where B is positive definite can be represented as TT^*, with $D_{T^*} = D_{A^*}$ and $T = A\beta$, $T^* = \beta A^*$, where

$\beta^2 = B$.

For confirmation and detailed discussion of statements e) and f) see, for example, Gel´fand and Shilov, Vol. IV [12], Chapter 1. Examples of such operators are easy to find. The beam, plate, shell, operators, the operators of linear elasticity, the Maxwell operators of electromegnetic field they are all of this type. See for specific examples, Volume I, Chapter 3 of this work.

For the discussion of the spectrum of operators of the type A^*A see the original article of Kato [39] or his monograph [27].

Chapter 4

Symmetry and Duality

4.1 Duality

Many forms of duality exist in mathematics. The Pontryagin duality, the duality between $L_1(-\infty, +\infty)$ functions and their Fourier transforms, the time-frequency duality in electrical engineering, the point-line duality in classical geometry, etc.

In 1964 B. Noble introduced the so called complementary variational principles based on the duality between linear operators acting between Hilbert spaces and their adjoints. (See [2],[1] for the original notes of Noble and [3] for a more comprehensive exposition.) An extensive exposition of the ideas of Noble and Sewell, outlining a wide range of applications of the complementary variational principles is given in the Noble and Sewell paper [4].

Development of Noble's theory on duality of operators in Banach (or even Fréchet spaces) can be found in papers of I. Herrera ([10], 1974), Herrera and Bielak ([13], 1976), and subsequent papers such as Collins [11], enlarging on their ideas. Tonti observed that potentiality of an operator is essential to the formulation of variational principles for nonlinear operators ([12]). Herrera observed that the symmetry condition for potential property of an operator in a linear vector space does not require an inner product structure, or even a norm to be defined in that space. Even without the inner product, adjoint of an operator may be defined by either completing some abstract diagram in some cases, or for example, by some integration by parts formulas.

Many arguments of this type rely on convexity and saddle-like behavior of corresponding functionals.

The problem of invariance of the saddle property of a functional under the action of some groups of transformations has not been properly researched at this time and only some sketchy results are available. In this respect the synge hypercircle ideas have not been fully incorporated into the techniques now available to engineers. The main idea related to the hypercircle approach is a decomposition of the linear vector space on which a functional is defined into two subspaces such that the functional is convex on one and concave on the other. Developments of this type have been originated by Ishmael Herrera and by Herrera and Sewell([10], [9]).

However the functional analytic approach to duality for generalized Hamiltonian systems and various generalizations discussed above have never been implemented to reproduce the geometric theories found in modern texts on Hamiltonian and Lagrangian mechanics.

To the best of our knowledge this is the first modest effort to connect these seemingly disjoint developments.

4.2 Hamiltonian mechanics.

For a detailed account of Hamiltonian classical and quantum mechanics read for example [17], [18] or [19]. The classical theories can be derived from our presentation, that follows roughly Korn and Noble ideas of functional analytic duality, by identifying the linear operators A and A* with differential operators $\frac{d}{dt}$ and $\frac{-d}{dt}$, respectively. There are several advantages of a more general scheme presented here. It also incorporates several ideas of control theory and optimization in a single postulation.

Let us start our discussion in a Hilbert space setting . Let H_1, H_2 be Hilbert spaces, A-a linear operator : $H_1 \to H_2$, with domain of A dense in H_1, A* a formal adjoint of A mapping H_2 into H_1, or H_2 into H_1^{-1} .

The equation of state for the system is given by:

$$Au = \mathcal{A}^* \mathcal{A} \underset{\sim}{u} + w(\underset{\sim}{x})\ \underset{\sim}{u} = q \tag{4.1}$$

$$u \in H_1(\Omega),\ q \ \in H_1^{-1}\ (\Omega),\ w(x) \in C^\infty(\Omega),$$

$$\mathcal{A}: \ H_1 \to H_2,\ \mathcal{A}^*: \ H_2 \to H_1 .$$

Ω is the domain of the function (vector function) $\underset{\sim}{u}$, $\Omega \subset \mathbb{R}^n$.

Denoting by $\underset{\sim}{p} = \mathcal{A}\underset{\sim}{u} \in H_2$,

we introduce the Lagrangian

$$L(u,\ \mathcal{A}\,u,\ \underset{\sim}{x}) = \tfrac{1}{2} < \mathcal{A}u,\ \mathcal{A}u>_{H_2(\Omega)}$$

$$+ \frac{w}{2} <u,u>_{H_1(\Omega)} - <q,u>_{H_1(\Omega)}\ , \tag{4.2}$$

if $q_1 \in H^1 \subset H^{-1}$.

If boundary conditions are assigned to u on $\partial\Omega$,

$$u_{\partial\Omega} = \phi(\underset{\sim}{x})\big|_{\underset{\sim}{x}\in\partial\Omega},$$

we may modify the Lagrangian to take care of the boundary conditions. Specifically, if we can introduce an inner product $[\ \ ,\ \ ]_{\partial\Omega}$ and a Hilbert space $H_3(\partial\Omega)$, of $L_2(\partial\Omega)$ functions, so that a suitable functional $\Gamma(\underset{\sim}{u}, \mathcal{A}\underset{\sim}{u}, \underset{\sim}{x})$ is assigned to $\partial\Omega$.

We shall delay the details of this technique. For the time being let us consider the case of natural boundary conditions (that is, we simply ignore them) and examine some properties of the Lagrangian $\mathcal{L}$ and of the corresponding Hamiltonian H, where a heuristic definition of H is given by

$$H = < u, \alpha^* p > - L(u, \alpha u) = < \alpha u,p> - L. \quad (4.2^a)$$

Here p is obtained by the Legendre transformation

$p = \frac{\partial L}{\partial (\alpha u)}$. Assumptions of smoothness, of the existence of the Fréchet derivative $\partial L/\partial(\alpha u)$, regularity of domain, etc., were made surreptitiously, and should be restated properly. However, this project can be carried out in an appendix. In some problems in engineering and physics the proper functional analytic setting can not be decided upon until all details of the expected physical behavior are known. For this reason physicists can get away with loose discussion of operators, not giving any specific details of what exactly do they operate on. In engineering problems it is easier to locate properly the vector spaces appropriate to the problem, but even here some nasty surprises are in store, and a very precise description of the domains, codomains and ranges of these operators may turn out to be futile when a solution of a specific physical problem defies the a - priori selected functional analytic setting. However, in presenting the main ideas of this work it is convenient to assume a Hilbert space setting. Until stated otherwise, all functions are vectors in a Hilbert space, all operators map Hilbert spaces into other Hilbert spaces, and all operators are linear and their domains are dense in the space.

Let us duplicate some of the classical arguments in the Hilbert space setting. We commence with the canonical system

$$(4.3^a) \quad \frac{\partial H}{\partial \underset{\sim}{u}} = A^* \underset{\sim}{p} \in H_1(\Omega) ,$$

$$(4.3^b) \quad \frac{\partial H}{\partial \underset{\sim}{p}} = A \underset{\sim}{u} \in H_2(\Omega)$$

where H $(u,p,\underset{\sim}{x})$ is the Hamiltonian.

Let τ be a parameter assuming values on R_+, so that we have

$$u = u(\tau, x)$$

$$p = p(\tau, x)$$

$$A = A(\tau)$$

$$A^* = A^*(\tau) \ .$$

If we assume that the state of a system with specific design $\tau = \hat{\tau} \in R_+$ is given by a solution of a (differential, differential-algebraic,...) equation $A^* Au = q$, with q independent of τ, we can define admissible states as a curve $\tau \to u(\tau, x)$ in $H_1(\Omega)$. The generalized canonical system of Hamilton corresponds to a critical point of the following functional

$$\mathcal{L} = \langle Au, p\rangle_{H_2} - \tfrac{1}{2}\langle p,p\rangle_{H_2} + \langle u, A^* p\rangle_{H_1} - \langle u,q\rangle_{H_1} .$$

The Fréchet derivatives of $\mathcal{L}$ are

$$\text{(4.4)} \qquad \begin{aligned} \mathcal{L}_u &= A^* p - q \\ \mathcal{L}_p &= Au - p. \end{aligned}$$

They vanish if $\left.\begin{aligned} p &= Au \\ q &= A^* p \end{aligned}\right\}$,

and if $\left.\begin{aligned} \frac{\partial H}{\partial u} &= A^* p \\ \frac{\partial H}{\partial p} &= Au \end{aligned}\right\}$

Suppose that along the orbit of τ, H has stationary

behavior at $\tau = \tau_o$.

Let us denote by U the vector of dependent variables

$$U = \begin{bmatrix} p \\ u \\ \tau \end{bmatrix} . \quad \text{Then} \quad U_\tau \equiv \frac{\partial U}{\partial \tau} = \begin{bmatrix} p_\tau \\ u_\tau \\ 1 \end{bmatrix} .$$

The evaluation map $U_\tau \big|_{\tau=\tau_o}$ gives us a triple of

real numbers $\begin{bmatrix} p_{\tau_o} \\ u_{\tau_o} \\ 1 \end{bmatrix}$.

Suppose that the total derivative of the Hamiltonian $D_\tau H = 0$ at the point $\tau = \tau_o$ of the orbit of τ is computed:

$$D_\tau H = \frac{\partial H}{\partial p} p_\tau + \frac{\partial H}{\partial u} u_\tau + \frac{\partial H}{\partial \tau} .$$

Therefore,

$$<D_\tau H, 1> = < \frac{\partial H}{\partial p}, p_\tau > + <\frac{\partial H}{\partial u}, u_\tau > + <\frac{\partial H}{\partial \tau} , 1 >,$$

and

$$\lim_{\tau \to \tau_o} \{<Au, p_\tau> + <A^* p, u_\tau> + <\frac{\partial H}{\partial \tau}, 1 >\} = 0 . \quad (4.8)$$

We define a Hilbert space H with the inner product of vectors $(u_1, p_1, f_1) = \mathbf{V}_1$, $(u_2, p_2, f_2) = \mathbf{V}_2$, $(u_1, u_2) \in H_1$, $(p_1, p_2) \in H_2$, $(f_1, f_2) \in L_2$, given by: $<u_1, u_2>_{H_1} + <p_1, p_2>_{H_2} + <f_1, f_2>_{L_2}$ $= \{ \mathbf{V}_1, \mathbf{V}_2\}$. Thus, we have the orthogonality criterion (4.8) for optimality at $\tau = \tau_o$

$$\lim_{\tau \to \tau_o} \{<(Au, A^* p, \frac{\partial H}{\partial \tau}), U_\tau>\} = 0 . \qquad (4.5)$$

We assume that all derivatives (displayed above) do exist.

Example. The Legendre-Germaine hypothesis of thin plate theory result in the following differential equation for the deflection function w(x,y,t):

$$[D(x, y) \nabla^2(\nabla^2 W) + 2\frac{\partial D(x, y)}{\partial x}\frac{\partial}{\partial x}(\nabla^2 W) + 2\frac{\partial D(x, y)}{\partial y}\frac{\partial}{\partial y}(\nabla^2 W) + \nabla^2(D(x, y))\nabla^2 W]$$

$$-(1-\nu)\left[\frac{\partial^2 W}{\partial x^2}\frac{\partial^2 D(x, y)}{\partial y^2} - 2\frac{\partial^2 W}{\partial x \partial y}\frac{\partial^2 D(x, y)}{\partial x \partial y} + \frac{\partial^2 W}{\partial y^2}\frac{\partial^2 D(x, y)}{\partial x^2}\right] = -\frac{\gamma\, h(x, y)}{g}\frac{\partial^2 W}{\partial t^2} + p(x, y),$$

$D(x, y)$ and $p(x, y)$ are known functions. On $\partial\Omega$ $W(x, y)$ obeys some physically motivated conditions which may be $W = 0,\ M_n = D\left\{\frac{\partial^2 W}{\partial n^2} + \nu\left(\frac{\partial^2 W}{\partial s^2} + \frac{\partial \psi}{\partial s}\frac{\partial W}{\partial n}\right)\right\} = 0$

where ψ is the angle which the boundary forms with the direction of the x axis, n denotes the direction of the outer normal to the boundary, while $\partial/\partial s$ denotes differentiation along the arc length of the boundary $\partial\Omega$.

It was shown in the Volume I of this work (pages 233-235) how this can be reduced to the form (4.3).

This system can be rewritten in a symbolic form(4.6), if $W_o(x) = e^{i\omega t}\, w(x,t)$, $M_o(x)\ e^{i\omega t}\, M(x,t)$,

$$\frac{\partial W_o}{\partial M_o} = -N^{-1} M_o = AV_o , \qquad (4.6^a)$$

$$\frac{\partial W_o}{\partial V_o} = V_o = A^* M_o , \qquad (4.6^b)$$

where $A = T\ \text{grad}$, $A^* = \text{div} \cdot T^*$.

$$T = \begin{bmatrix} \partial/\partial x & 0 \\ 0 & \partial/\partial y \\ \partial/\partial y & 0 \\ 0 & \frac{\partial}{\partial x} \end{bmatrix} \qquad (4.7)$$

and T^* is the transpose of T .

Our orthogonality criterion for optimality of a parameter τ becomes

$$\lim_{\tau\to\tau_o} \{ <AV_o, M_{o_\tau}> + <A^* M_o, V_{o_\tau}> + <\mathcal{W}_{o_\tau}> \} = 0. \quad (4.8)$$

A similar argument applied to the Euler-Bernoulli's model of a beam in static bending results in the equality

$$q\mathcal{W}_h + \tfrac{1}{2} EI_h (\mathcal{W}_{xx})^2 = 0 , \quad (4.9)$$

allowing us to compute $\mathcal{W}_h$.

The Fréchet derivatives displayed here have the same meaning as before. (See appendix 1 to volume 1, or the discussion in Chapters 2 & 3 of volume 1.

For example, the definition of $\mathcal{W}_h$, I_h is given as follows.

$$\mathcal{W}_h(x)\Big|_{h=\bar{h}} = \lim_{\varepsilon\to 0} \frac{\mathcal{W}(x,\bar{h}(x) + \varepsilon\eta(x)) - \mathcal{W}(x,\bar{h}(x))}{\varepsilon},$$

and

$$I_h\Big|_{h=\bar{h}} = \lim_{\varepsilon\to 0} \frac{I(x,\bar{h}(x) + \varepsilon\eta(x)) - I(x,\bar{h}(x))}{\varepsilon}$$

exist, provided $\mathcal{W}_h(x,\eta(x))\Big|_{h=\bar{h}}$ and

$I_h(x,\eta(x))\Big|_{h=\bar{h}}$ are both independent of $\eta(x)$,

where $\eta(x)$, $x \in [0, \ell]$, is an arbitrary admissible function, with $||\eta(x)||_{L_\infty} < 1$.

Comment. The problem of sensitivity is in effect a problem complementing the well-posedness problem of a partial differential equation describing the state of the system. The well-posedness demands that small changes in the values of the parameters produce only small changes in the solutions. But such situation does not permit one to improve significantly a value of the cost functional by making small adjustments in the values of parameters of the system which depend on the design.

The improvements in the design are most dramatic if the system is very sensitive to such design changes.

In fact, if after some optimizing steps the sensitivity of the system to design changes is very low, one can accept the design as optimal and terminate a successive approximation procedure.

Roughly speaking, the equilibrium is stable if small changes in the form of the function $f(x,t)$ or in initial conditions assigned to (4.1) cause the system to return to an equilibrium. Similarly the design of a system is stable if the sensitivity of the state with respect to the design changes is "small".

This topic will be discussed in much greater detail later on. For the present let us offer an outline of a proof concerning the existence of a continuous derivative of the state of the system with respect to the design for engineering systems whose behavior is given by equation of the Hamilton-Kato type:

$$AA^* u = q.$$

Essentially, we are concerned with the behavior of the "variational" bilinear functional $\langle AA^* u, v\rangle$, where for sake of simplicity we assume that $\langle \ , \ \rangle$ is a simple $L_2(\Omega)$ product. Several papers were written on linear equations involving operators of the form AA^* and variants of that form. An important paper of J.J. Moreau containing the so called "two cones theorem" and subsequent papers concerning Hilbert space decomposition for a class of problems involving operators of the form AA^* were published in the 60-s and 70-s. See [5], [6], [7], [8].

For example the Dirichlet, Neumann and Robin problems for the Laplace operator are generated by the operators $A = \text{div}$, $A^* = \text{grad}$. A possible domain D_A of A is a subset of the real Hilbert space $L_2^{(n)}(\Omega)$, $\Omega \subseteq \mathbb{R}^n$, whose elements are n-dimensional vectors, that is ordered n-tuples of real numbers. The inner product in $L_2^{(n)}(\Omega)$ is given by

$$< \underset{\sim}{u}, \underset{\sim}{u} >_{L_2^{(n)}} = \int_\Omega \left(\sum_{i=1}^{n} u_i v_i\right) d\underset{\sim}{x}.$$

Let D_0 denote a set of all functions in domain $D_A \subsetneq L_2^{(n)}(\Omega)$ that vanish identically on $\partial\Omega$, and D_1 of all vector functions in D_A whose component normal to the boundary $\partial\Omega$ vanishes identically on $\partial\Omega$.

The operator $A^* = \text{grad}$ maps $D_{A^*} \subsetneq L_2(\Omega)$ $(= L_2^{(\perp)}(\Omega))$ into D_A. The inner product in $L_2(\Omega)$ **is as usual** $\langle f,g\rangle_{L_2} = \int_\Omega (f \cdot g)\, dx$.

Depending on boundary conditions assigned on $\partial\Omega$ we can choose appropriate subdomains of D_A and D_{A^*} respectively that essentially enforce these boundary conditions.

These were extremely simple cases of of problems with operators of the form AA^* defined on "reasonable" domains. An obvious decomposition of the domain of AA^* uses a basic theorem of functional analysis that the range of A is the orthogonal complement of the null space of A^*. For the sake of completeness let us repeat a standard argument. Suppose that $v \in N_{A^*}$ (the null space of A^*). Then, obviously $A^* v = 0$, and $\langle u, A^* v\rangle = 0$ for any u in the domain of A, since $\langle u, A^* v \rangle = \langle A u, v \rangle = \langle w,v\rangle$ **for any w in the** range of A. Thus, any w in the range of A is orthogonal to any v in the null space of A^*.

For some consequences of this statement see for example Dunford and Schwartz [13].

4.3 Hilbert space duality for operator equations with essential boundary conditions.

We recall that boundary value conditions assigned to variational problems are called natural if the solution of the problem with no boundary conditions assigned to it, satisfies the natural conditions anyway. Roughly speaking they do not affect the solution of the variational problem. Boundary conditions that are not natural are called essential. The natural conditions can be ignored, the essential conditions cannot be ignored in solving a variational problem.

We can accomplish the incorporation of essential conditions by extending the domain of the operator. For example, the linear operator equation

$Tu = f \quad , \quad T : H_1 \to H_2 \quad$, where H_1 and H_2

are real Hilbert spaces, and the vectors $u \in H_1$, $f \in H_2$ are functions defined on $\Omega_1 \subset R^m$ and $\Omega_2 \subset R^n$, respectively. The domain of T is dense in H_1 (so that the adjoint operator T* can be uniquely defined), and it is closed. We can define H_2, so that the domain of T* is all of H_2 and then the operator T* is closed, no matter whether T was closed or not. The proof of that statement is almost trivial, but it is included for the sake of completeness. Let u be a vector in the domain of T. Then for any $g \in H_2$ we can define

$\langle Tu,g\rangle_{H_2} = \langle f,g\rangle_{H_2}$, for any $u \in D_T$ and

$\langle T^*g, u\rangle_{H_1} = \langle g,f\rangle_{H_2}$.

If u is not in the domain of T, we define

$\langle Tu,g\rangle = \langle u, T^*g\rangle$ in the style of Sobolёv "weak" definition (see [14]). The Riesz theorem implies that there exists a vector $h \in H_1$, such that $\langle Tu,g\rangle_{H_2} = \langle u,h\rangle_{H_1}$. We set $T^*g = h$, obtaining the desired equality $\langle Tu,g\rangle = \langle u,T^*g\rangle$, and showing that any $g \in H_2$ will do if $u \in D_T$.

The simplest example is the Dirichlet problem.

$$(*) \quad \begin{cases} -\Delta u = f & \text{in } \Omega \subset \mathbb{R}^2 \\ u \equiv 0 & \text{on } \partial\Omega, \end{cases}$$

where f is once differentiable and bounded in Ω (of course we could use slightly weaker assumptions!), while Ω is a bounded, open, a simply connected region in $\mathbb{R}^2$, with Lyapunov boundary $\partial\Omega$. (See [14] for definition, or volume 1 of these notes) Δ denotes Laplace's operator. We can select as our Hilbert space H_1, the space $H_0^1(\Omega)$ of the functions which have at least one weak derivative in Ω, and vanish identically on $\partial\Omega$.

This problem has a unique solution $u_o \in H_0^1(\Omega)$. We decompose the problem (*) into a system

$$(**) \quad \begin{cases} \left.\begin{array}{l} \text{grad } u = \vec{v} \\ \text{div}(\vec{v}) = f \end{array}\right\} \text{ in } \Omega, \\ u \equiv 0 \quad \text{on } \partial\Omega. \end{cases}$$

Now, let us consider imbedding the space $H_0^1(\Omega)$ in a larger space $H^1(\Omega)$ with the inner product $\langle \xi,\eta\rangle_{H^1} = \int_\Omega \{\xi(\underset{\sim}{x})\eta(\underset{\sim}{x}) + \sum_{i=1}^{2} (\frac{\partial\xi}{\partial x_i} \frac{\partial\eta}{\partial x_i})\} d\underset{\sim}{x}$

with ξ,η satisfying some (not necessarily homogeneous) boundary conditions on $\partial\Omega$. Let $\tilde{u}$ denote the solution of the Dirichlet problem in the larger space: $H^1(\Omega)$

Let T_o denote the restriction of operator T to

the space H_0^1. The space H_0^1 is a subspace of H^1. Therefore any vector $u \in H^1$ may be decomposed. $u = u_o + \eta$ where $u_o \in H_o^1$ and η in the orthogonal complement of H_o^1 in H^1. We note that $\|\text{grad } u|^2_{H^1} = 0$ implies that $u \equiv$ constant. But if $u \equiv$ constant and $u \equiv 0$ on $\partial\Omega$ then $u \equiv 0$ in Ω. Thus, the null space of the operator $T = \text{grad}$ in $H_o^1(\Omega)$ consists only of the null vector that is of the function $u \equiv 0$. (We recall that discontinuous functions were not admitted in the domain of T.) Since $D_{T_o} \subseteq D_T$ we have $D_{T^*} \subseteq D_{T_o^*}$, which can be written concisely $T^* \subseteq T_o^*$. Recognizing T^* as the divergence operator, we can solve the problem (**) in H_0^1: $T_o^* T_o u_o = f$, which is equivalent to the problem

(***) $T^* T_o u_o = f$, in H_o^1.

The null space of the operator T_o consists only of the zero vector. Thus, problem (***) (with an admissible function $f(\underset{\sim}{x})$) has a unique solution u_o in H_o^1.

Now considering the problem in H^1 with $u = \phi(\underset{\sim}{x})$ on $\partial\Omega$, $\phi(\underset{\sim}{x}) \not\equiv 0$, we only need to find a function $g(\underset{\sim}{x})$ such that $g \in D_T$ (therefore $g \in D_{T_o}$), $T^*Tg = 0$ in Ω and $g(\underset{\sim}{x}) = \phi(\underset{\sim}{x})$ on $\partial\Omega$.

We observe that the decomposition of H^1 into

into H_o^1 and $(H_o^1)^{\perp}$ allows an easy norm estimate:
$||\xi||_{H^1} \geq ||\xi||_{H_o^{\perp}}$ for any $\xi \in H^1$.

The bounds for approximate solutions may be now obtained from the Hilbert space parallelogram law.

4.4 Other suitable functional analytic background for engineering problems.

The Hilbert space setting for engineering problems comes naturally. For example the finite total energy criterion for the classical(linear) model of beam, plate or shell theory is given by the quadratic form inequality. For example,the finiteness of strain energy for the beam is given by the inequality

$$\int_o^{\ell} EI(x)(y_{xx})^2 \, dx \leq M \quad \text{for some real number } M.$$

This is equivalent to existence (or finiteness) of the inner product <y,w> given by

$$\langle y,w \rangle = \int_o^{\ell} EI(x)(w_{xx}y_{xx}) \, dx. \qquad (4.10)$$

This dictates the functional analytic setting for the problem in Sobolev space H^2:

$$\langle y,w \rangle = \int_o^{\ell} \sum_{i=0}^{2} (D^i w \cdot D^i y) \, dx , \quad D^i = \frac{\partial^i}{\partial x^i} , \qquad (4.11)$$

in this one-dimensional problem. Similarly the "variational form" for the linear plate theory is given by

$$\langle w,\psi \rangle_{\Omega} = \tfrac{1}{2} \iint_{\Omega} \{\nabla^2[D(x,y)\nabla^2 w_o(x,y)] \cdot \psi(x,y) -$$

$$(1 - \nu) \lozenge^4 (D(x,y), w_o(x,y)) \cdot \psi(x,y)\} \, dx \, dy \qquad (4.12)$$

with $w \equiv \frac{\partial w}{\partial n} \equiv 0$ on $\partial\Omega$

(built in edge condition along the entire boundary) or we can introduce a complementary energy product.

$$\langle w,\psi\rangle_\Omega = \tfrac{1}{2} \iint_\Omega D(x,y)\Big[\frac{\partial^2 w}{\partial x^2}\frac{\partial^2 \psi}{\partial x^2} + \frac{\partial^2 w}{\partial y^2}\frac{\partial^2 \psi}{\partial y^2}$$

$$+ \nu \Big(\frac{\partial^2 w}{\partial x^2}\frac{\partial^2 \psi}{\partial y^2} + \frac{\partial^2 w}{\partial y^2}\frac{\partial^2 \psi}{\partial x^2}\Big) + 2\cdot(1-\nu)\Big(\frac{\partial^2 w}{\partial x\partial y}\frac{\partial^2 \psi}{\partial x\partial y}\Big)\Big]$$

$$dxdy. \qquad (4.13)$$

In dynamic problems we add the term

$$\tfrac{1}{2} \iint_\Omega \Big[\rho(x,y)\Big(\frac{\partial w}{\partial t}\frac{\partial \psi}{\partial t}\Big)\Big]\,dxdy \qquad (4.13^a)$$

in the definition of $\langle w,\psi\rangle$.

A different view of elastodynamics proposes the dependence on time of the terms in Newton's second law as applied to constitutive properties of solids. For example, the Newton's second law relates the stress component derivatives to the forces that cause them by equations of motion

$\tau_{ij,j} + f_i = \rho\ddot{u}_i$, (summation convention is used) if certain linearalizing assumptions are made; (see Novozhilov [15] for an explanation of this statement; and see [16].)

If we also postulate the Hooke's law as the constitutive equation

$$\tau_{ij} = C_{ijk\ell}\,\varepsilon_{k\ell} \qquad (4.13^b)$$

and the compatibility equations (4.13^c)

$\operatorname{curl}(\operatorname{grad} \tilde{u}) = 0$,

we have the equations of classical elasticity.

The viscoelastic materials obey (instead of Hooke's law) constitutive equations relating the stresses to the time rates.

Such law may be of the form

$$\tau_{ij} = \int_{-\infty}^{+} G_{ijk\ell}\,(x,\ t-\tau)\cdot\dot{\varepsilon}_{k\ell}\,(x,\tau)\,d\tau,$$

with $G_{ijk\ell} = G_{jik\ell} = G_{k\ell ij} = G_{k\ell ji}$,

and $\varepsilon_{ij} = \frac{1}{2}(u_{i,j} + u_{j,i})$,

(thus $\varepsilon_{ij} = \varepsilon_{ji}$).

The initial conditions at t = 0 are given by

$\underset{\sim}{u}(o) = u_o$,

$\underset{\approx}{\varepsilon}(o) = \underset{\approx}{\varepsilon}_o$, $\underset{\approx}{\tau}(o) = \underset{\approx}{\tau}_o$.

Defining:

$\sigma_{ij} = \int_{-\infty}^{t} \dot{G}_{ijk\ell}(x,t+\tau)\, \varepsilon_{ij_o}(x,\tau)\, d\tau$, the following constitutive equation can be adopted to model a viscoelastic solid

$$\bar{\sigma}_{ij} = \sigma_{ij} + \frac{d}{dt}(G_{ijk\ell} * \varepsilon_{k\ell}) .$$

Also, the Tierstein's equation of motion for a piezoelastic solid is given by a similar form

$$(4.14) \quad g * \sigma_{ij,j} = b_i + \rho u_i$$

where g is the memory Kernel and * is the usual convolution product. Similar relations have been established in various areas of continuum mechanics. In thermodynamic processes one can associate the temperature of the system with an average of kinetic energy of the molecules, or one can introduce it as in independent variable, perhaps, later regretting this simple assumption. Suppose that the kinetic energy is given by a general quadratic relation in real displacement (generalized) variables u^i and internal variables q^i of the molecules that is of the form:

$$(4.14^a) \quad T_{1,2} = \tfrac{1}{2} \sum_{i,j} \int_{\Omega} \{\rho(\underset{\sim}{x})\, \dot{q}_1^i\, q_2^j\}\, dx \; +$$

$$\tfrac{1}{2} \sum_{\alpha\beta\gamma\delta} \int_{\Omega} (\dot{C}_{\alpha\beta_1}\, \dot{C}_{\gamma\delta_2})\, dx\, ,$$

where C_{ij} are strain components.

We can also introduce a memory kernel g relating stresses to strains in a constitutive equation.

$$(4.14^b) \quad \tau_{ij} = \int_{-\infty}^{T} g_{ijk\ell}\, (\underset{\sim}{x},\, t - \tau)\, C_{k\ell}(x,\tau)\, d\tau\, .$$

However, even theoretical linear elastodynamics may be rewritten as a convolution kernel. Integrating twice the equilibrium equations we obtain

$$t * \tau_{ij,j}(\underset{\sim}{x},t) + t * f_i = \rho u_i,\; i = 1,2,3\, .$$

If the body force $\underset{\sim}{f}\ (\underset{\sim}{x})$ is not a function of time, then $t * f$ is a quadratic function of time. (Note: The linear form of equilibrium equations is assumed in our statement above, i.e.

$$\tau_{ij,j} + f_i = \rho\ddot{u}\ .)$$

As was shown by I. Herrera and by Morton Gurtin, the basic duality theory along the lines of Noble and Sewell can be duplicated step by step by replacing inner products by the appropriate convolution algebra. Connection between duality in the sense of Noble and existence of Noether invariants can be established, but it is not easy.

The simplest interpretation of the Noble's duality is obtained in the classical Hamiltonian systems when we identify in the "dual" equations (4.3^a), (4.3^b) the operator A with $\frac{d}{dt}$ and A* with $\frac{-d}{dt}$. Then this duality becomes a simple integration by parts principle.

There is a rather humorous story circulated in the mathematical community about Solomon Lefschetz's visit to France when he walked into Schwartz's office demanding a definition of weak derivatives and of distributions. After hearing it he looked incredlous saying: "But this is just integration by parts!" When Schwartz replied "Exactly! That's what it is!", he walked out saying "I only hope, young man, that you have worked on some more serious mathematics than this."

In fact, the entire areas of Noble and even Lefschetz and Pontryagin duality can be regarded as abstract generalizations of the integration by parts idea. Alternatively the integration by parts can be regarded as a specialized example of some deep principles of symmetry pervading all mathematics. It is historically significant that many ideas of Cartan originated from his interest in the mechanics of rigid bodies. The variational formulation of Hamiltonian mechanics can be associated with two functionals.

$$(4.15^a)\quad K(t,q,p) = \int_0^T [H(t,q,p) - p\dot{\tilde{q}}\,]dt\ ,$$

$$(4.15^b)\quad J\,(t,q,p) = \int_0^T [H(t,q,p) + q\dot{p}]dt.$$

Natural boundary (or initial) conditions are assumed.

If we postulate a relation $p = \phi(x)q$, we can relate the behavior of functionals K and J given by (4.14).

$$I_1(q,p) = \int_0^T L(x,q,p)\ dt,$$ subject to a constraint

$\phi(x)\dot{q} - p = 0$, where $\phi(x)$ is positive. Assuming that the derivative of $\phi(x)\dot{q} - p$ does not vanish and a Lagrangian multiplier use is justified, the extremum of I_1 is replaced by the extremum of the functional:

$$I_2 = I_1 + \int_0^T [\lambda(\phi(x)\ \dot{q} - p)]dt = \int_0^T [L(x,q,p) +$$

$\lambda(\phi\dot{q} - p)]$ dt.

Integration by parts gives

$$(4.15^a) \quad \frac{\partial L}{\partial p} = -\lambda ,$$

$$(4.16^b) \quad \frac{\partial L}{\partial q} = \frac{d\lambda}{dt} .$$

and also

$$(4.17) \quad \dot{q} = p\,\phi^{-1}(x) .$$

Noble's representation of various physical phenomena consists in the following generalization.

The operator $\frac{d}{dt}$ is replaced by a suitable linear operator A and $-\frac{d}{dt}$ by A^*, which is either the "real" adjoint of A or only the formal adjoint of A, derived after ignoring some initial and/or boundary conditions.

More specifically A could be a matrix of linear differential or algebraic operators, it could be operator of convolution with the memory Kernel, or a "mixed" integro-differential operator.

The discovery of such general representation duplicating Hamiltonian mechanics around 1963 produced a massive outpouring of papers.

As with every theory that gives new perspectives to "thoroughly examined" problems in engineering or physics, Noble's informal presentation created a crowd of followers plus some reasonably justified claims that this point of view is not really new.

One could argue that this type of duality is in essence a more general exposition of Korn's duality or the Korn-Friedrichs' duality that has been known in theoretical elasticity. Other papers that pioneered this approach in certain restricted cases include H. Bateman's 1929-1930 papers [20], [21], the C.C. Lin and S.I. Rubinov paper [22], the paper of M. Shiffman [23] or the

Lush and Cherry ideas of [24]. These early applications dealt either with compressible fluid flow or the large deflection problems in elasticity. For example, let us consider a three-dimensional flow of a compressible fluid, satisfying four basic relations

$$\operatorname{div}(\rho \underset{\sim}{v}) = 0$$

$$\rho[(\underset{\sim}{v} \cdot \operatorname{grad})\, \underset{\sim}{v}] = -\operatorname{grad} p \,.$$

The equation of equilibrium becomes

$$f'(\rho) - \tfrac{1}{2} v^2 = 0 \,. \tag{4.23}$$

If we follow Noble's arguments, we can assemble all equations (4.20)-(4.23) and show how they fit into the generalized Hamiltonian system.

Introducing the velocity potential Φ, we have

$$\left.\begin{aligned} &(4.24^a) \quad \operatorname{grad} \Phi = \underset{\sim}{v} \\ &(4.24^b) \quad \operatorname{div}(\rho \underset{\sim}{v}) = 0. \end{aligned}\right\}$$

Denoting by $\underset{\sim}{q} = \rho \underset{\sim}{v}$,

we rewrite (4.24^a)

$$(4.25^a) \quad \frac{\partial \Phi}{\partial x} = \frac{q_1}{\rho}, \; \frac{\partial \Phi}{\partial y} = \frac{q_2}{\rho}, \; \frac{\partial \Phi}{\partial z} = \frac{q_3}{\rho},$$

q_1 denoting the x-component of $\underset{\sim}{q}$, etc.

$$\left.\begin{aligned} &(4.25^b) \quad \operatorname{div}(\underset{\sim}{q}) = 0, \\ &(4.25^c) \quad \rho^2 \tilde{f}'(\rho) = \tfrac{1}{2} Q^2, \end{aligned}\right\}$$

where $Q^2 = \left(\sum_{i=1}^{3} q_i^2 \right)$.

If we produce a Hamiltonian W such that

$$W(q) = W(q(\rho)) = \tilde{W}(\rho), \text{ and } \frac{\partial W}{\partial \underset{\sim}{q}} = (\rho^{-1}) \cdot \underset{\sim}{q}$$

then the system of equations (4.25) becomes

$$(4.26^a) \quad \frac{\partial \Phi}{\partial x_i} = \frac{\partial W}{\partial q_i}(\underset{\sim}{q})$$

$$(4.26^b) \quad \text{div}\,(\underset{\sim}{q}) = \frac{\partial W}{\partial \Phi} = 0$$

(W is not an explicit function of Φ).

We have four equations in five unknowns : ρ, p, and the three-dimensional vector $\underset{\sim}{\mathbf{v}}$. Generally p is not a function of ρ only. Flows in which $p = p(\rho)$ (for example $p = k\rho^\gamma$ where k is a constant depending on the initial conditions and γ is a physical constant) are called adiabatic, isentropic or homentropic. In the more general case $p = p(\rho, S)$ where S is the entropy of the fluid or gas is the remaining "constitutive" equation. The flow is called irrotational if

$$(4.20) \quad \text{curl}\, \underset{\sim}{\mathbf{v}} \equiv 0.$$

In that case

$$\underset{\sim}{\mathbf{v}} \times \text{curl}\, \underset{\sim}{\mathbf{v}} \equiv 0.$$

Since $(\underset{\sim}{\mathbf{v}} \cdot \text{grad})\, \underset{\sim}{\mathbf{v}} = \frac{1}{2}\, \text{grad}(\|\mathbf{v}\|^2) - \underset{\sim}{\mathbf{v}} \times \text{curl}\, \underset{\sim}{\mathbf{v}}$,

we have

$$(\mathbf{v} \cdot \text{grad})\, \mathbf{v} = \frac{1}{2}\, \text{grad}(\|\underset{\sim}{v}\|^2).$$

Denoting $\|\underset{\sim}{\mathbf{v}}\|$ by V, we have

$$(4.21) \quad \frac{1}{2}\, \text{grad}\,(V^2) + \frac{1}{\rho}\, \text{grad}\, p = 0 \cdot$$

Since $p = p(\rho)$,

we can follow Bateman's arguments and introduce a function $f(\rho)$ such that

$$(4.22) \quad p(\rho) = f(\rho) - \rho\, f'(\rho).$$

Thus $\frac{dp}{d\rho} = -\rho\, f''(\rho)$.

$$\frac{\partial \tilde{W}}{\partial \rho} = \frac{\partial W}{\partial \underset{\sim}{q}} \cdot \frac{\partial q}{\partial \rho} = \frac{1}{\rho} \quad (\underset{\sim}{q} \cdot \frac{\partial p}{\partial \rho}) = f'(\rho) + \frac{d}{dp}(\rho f'(\rho)).$$

We identify after some manipulation

$$\tilde{W} = f(\rho) + \rho\, f'(p) = f(\rho) + \tfrac{1}{2}\, \rho\, V^2.$$

We can identify the functionals I_1, I_2.

$$I_1 = \int_\Omega \tilde{W}(\rho)\, d\underset{\sim}{x} - \int_\Omega (\underset{\sim}{q}\cdot \text{grad}\, \Phi)\, d\underset{\sim}{x} = \int_\Omega p\, d\underset{\sim}{x}$$

$$= \hat{W} - \langle q, A\Phi \rangle .$$

$$I_2 = \int_\Omega \tilde{W}(\rho)\, dx - \int_\Omega (\text{div}\, \underset{\sim}{q} \cdot \Phi)\, dx$$

$$= \int_\Omega \tilde{W}(\rho)\, d\underset{\sim}{x} = \int_\Omega (f(\rho) + \tfrac{1}{2}\, \rho\, V^2)\, d\underset{\sim}{x}$$

$$= \int (p + \rho\, V^2)\, d\underset{\sim}{x} .$$

This is basically a rewriting of the results of Bateman and of Cherry and Lush. A more detailed analysis of these authors implies that the extrema of I_1 and I_2 occur only for the subsonic flows. For supersonic flows the signs are all wrong and only the stationary quality of these functionals can be deduced.

However, an original idea of the author presented in Chapter 5 of volume one (under the title of "different algebra") can be used to generate variational principles exactly in such cases when the "wrong" signs frustrate the classical variational formulation. So far it has not been applied to supersonic flows extending the Cherry-Lush type analysis.

4.5 Lagrangian Systems with velocity constraints

Let $\underset{\sim}{q} = \{q_1, q_2 \ldots q_n, q_{n+1}, \ldots q_m\}$ be a local coordinate system, that is a local isomorph of R^m. The dynamics of a mechanical system are specificed by the extremum of an action integral $\delta \int_0^T L \, dt = 0$, with the Lagrangian of the form

$L(\underset{\sim}{q}, \dot{\underset{\sim}{q}}, t)$. The coordinates $\underset{\sim}{q} = (q_1, q_2, \ldots, q_{n+1}, \ldots q_m)$ describe the state of the system. m - n constraints are given in the form

$$(4.23) \quad \dot{q}_j = \sum_{k=1}^{n} c_{jk}(\underset{\sim}{q}) \, q_k , \; j = n+1, \ldots m.$$

These are nonholonomic constraints of the "velocity type". As usual, we form the modified Lagrangian

$$(4.24) \quad \tilde{L} = L + \sum_{j=n+1}^{m} \mu_j [\dot{q}_j - \sum_{k=1}^{n} c_{jk}(\underset{\sim}{q}) \, \dot{q}_k],$$

where μ_j are Lagrange multipliers. $\frac{\partial \tilde{L}}{\partial \dot{q}_j} = \mu_j$ if $j > n$. (4.24).

Obviously, we form the Gateaux difference

$$\Delta \tilde{L} = \tilde{L}(\underset{\sim}{q} + \varepsilon \underset{\sim}{\eta}, \dot{\underset{\sim}{q}} + \varepsilon \dot{\underset{\sim}{\eta}}, t) - L(\underset{\sim}{q}, \underset{\sim}{q}, t)$$

and collect carefully first order terms in $\varepsilon\eta$.

Let $\hat{L}$ stand for $\tilde{L}$ with all $\dot{q}_j (j>n)$ eliminated from $\tilde{L}$ by making use of constraints (4.23). Vanishing of the first variation corresponds to the zero of

$$< \varepsilon\eta_i, (\frac{\partial \hat{L}}{\partial q_i} - \frac{d}{dt}(\frac{\partial \hat{L}}{\partial \dot{q}_i})) > - < \varepsilon\eta_i, (\sum_{k=1}^{n} \{ \sum_{j=n+1}^{m} \mu_j \cdot$$

$$\cdot\left(\frac{\partial c_{jk}}{\partial q_i} - \frac{\partial c_{ji}}{\partial q_k}\right)\}\dot{q}_k\rangle, \qquad i = 1,2,\ldots n.$$

The Euler-Lagrange equations for $\hat{L}$ become the Chaplygin's equations

$$(4.25)\quad \frac{\partial \hat{L}}{\partial q_i} - \frac{d}{dt}\left(\frac{\partial \hat{L}}{\partial \dot{q}_i}\right) - \sum_{k=1}^{n} \dot{q}_k\left(\sum_{j=n+1}^{m} \left\{\frac{\partial \hat{L}}{\partial \dot{q}_j}\left(\frac{\partial c_{jk}}{\partial q_i} - \frac{\partial c_{ji}}{\partial q_k}\right)\right\}\right) = 0$$

which, written in the "weak form" are

$$(4.25^a)\quad \langle\lambda_i, \left(\frac{\partial \hat{L}}{\partial q_i} - \frac{d}{dt}\frac{\partial \hat{L}}{\partial \dot{q}_i}\right)\rangle - \langle\lambda_i\{\sum_{k=1}^{n} \dot{q}_k\left(\sum_{j=n+1}^{m} \{\frac{\partial L}{\partial \dot{q}_j}\left(\frac{\partial c_{jk}(\tilde{q})}{\partial q_i} - \frac{\partial c_{ji}(\tilde{q})}{\partial q_k}\right)\}\right\}\rangle = 0, \quad i=1,2,\ldots n, \text{ with } \lambda_i=\lambda_i(t).$$

There is a problem in formulating a Noble-type duality for equations (4.25) or (4.25^a).

It has been shown by some authors that it cannot be done. See, for example, Sumbatov[25] who gave a specific counterexample, and offered the following theorem.

In general there does not exist a coordinate change, such that Chaplygin's equations can be (locally) reduced to the Euler-Lagrange form in the new coordinate atlas.

Consequently, there is no Hamiltonian such that Hamiltonian-type duality can be obtained directly from the Chaplygin's equations.

We should qualify that statement. Sumbatov's counterexample, concerning the rigid body equations, implies(in general) the impossibility of a coordinate change that accomplishes such transformation, which does produce an appropriate Lagrangian and Hamiltonian for the rigid body system. It does not say that such system could not be imbedded in higher dimensional system for which such transformation is possible, and there-

fore the Hamiltonian duality can be realized in such imbedding. This seems to be a good research topic for an ambitious graduate student. The author conjectures a positive answer to this question of existence.
The problems of imbedding an otherwise unsolvable problem in a higher dimensional space has a long history, going to the 19th Century. Some spectacular results or conjectures are now famous in mathematical physics. Probably, the most notable were the Kaluza-Klein attempts to imbed quantum theories in a continuous theory by postulating the existence of paths that are continuous in higher dimensions but puncture the hyperplane of our time·space at isolated points.
The author "played games" with imbedding continuum mechanical phenomena by varying some physical constants (see volume 1, pages 284-286). A fundamental theorem concerning imbeddings in a higher dimensional Hilbert spaces that restore normality to linear operators was proved by Halmos. It appears to be a crucial theoretical result in studying spectral theory and the von Neumann representation in continuum mechanics.

However, this topic would take us too far afield and away from the main topics of this volume. For in-depth study of stability, causality and higher dimensional imbedding based on spectral decomposition see the monograph of R. Saeks[22].
The whole topic of causality has been deliberately avoided in our discussion, but it is of great importance in interpreting "physically" some of the results of the mathematical models of the "real world" phenomena. It suffices to mention the fact that non-causal models may acquire causality or anticausality(the future determine the past!) by a suitable imbedding in a higher dimensional model. The question of causality is closely related to normality of operators. (Note: Operator A is normal if A commutes with its adjoint.) A fundamental result of Halmos has resolved the question of imbedding operators in a larger space to attain normality.

4.6 Other forms of duality

There are many seemingly different aspects of the concept of duality. The early observations of H. Hertz in his deep treatise on mechanics [28] identified some fundamental difficulties in defining generalized forces acting on a mechanical system. According to Hertz, in a mathematical model of a mechanical system forces represent linear bounded maps from the space of generalized displacements into real (or complex) numbers. Various forms of duality have been handled separately in mathematical literature, in physics and in engineering. Extensive literature exists on different topics intimately related to abstract formulation of duality, such as the duality between a group and the group characters (see, for example L.V. Pontryagin's monograph Topological groups [27]). In a Hilbert space setting the duality of function spaces has to be emphasized as a physical concept, since mathematically a separable Hilbert space H and its topological dual are isomorphic. Only when the mapping take place between different Hilbert spaces does the duality of physical concepts match the duality of mathematical symbolism.

It is easier to offer specific examples than to explain the difficulties arising from the mathematical convenience, where H and H* are identified and no bookkeeping is necessary to separate functions in H from functions in H*.

Let us consider a real $L_2(\Omega)$ space, where Ω is an open region in $\mathbb{R}^2$. There is no way of distinguishing elements in $L_2(\Omega)$ from those in the dual and some of the most attractive aspects of Hilbert space theories result from this familiar identification of space with its dual. Unfortunately, such convenient identification cannot be made if elements of $L_2(\Omega)$ (or other Hilbert space) represent physical quantities.

The mathematical analysis generally manages to ignore the physical dimensional analysis. The bra-ket notation of Dirac preserves the separate identity of dual spaces, such as generalized displacements, and corresponding generalized forces, entropy and temperature, voltages and currents in the multiloop networks.

In general, mathematicians have been carefully avoiding the concept of a physical dimension. Identifying specifically concepts that relate physically to distances and denoting very carefully that a specific differential equation relates to each other only physical concepts of spatial distance never appealed to our graduate students. If an identical equation would also relate to each other the relevant "forces" acting on the system, a mathematician would ignore the physics and say "a solution is a solution", and who cares what is the physical interpretation of that solution, except if your boss happens to be a physicist or engineer, who cares about the physical dimensions of the quantities.

With the advent of functional analysis, the literature on duality between abstract vector spaces is extremely large and pertinent introduction can be found in many textbooks on functional analysis, for example in [29].

In duality between operators acting on an inner product vector space and their adjoints has been discussed informally in volume 1 of this work. For an extensive theory of this duality in Hilbert spaces see [29],[30].

The physically motivated duality arising in dimensional analysis is rarely discussed in mathematical texts.

Some very interesting work extending the so called "Pi theorem" of Buckingham may be found in the works of L.I. Sedov and his collaborators, in the Polish school (Drobot, Kasprzak, Lysik and their students) and in the United States(Birkhoff, Bridgeman, Na and Hansen, Cheng and others). There is an intimate relation between the concepts of similarity (and self-similarity) and the dimensional analysis.

The multiparameter groups of transformations that preserve similarity relations form a Lie algebra. Appropriate contact transformations may be developed to systemize this procedure. This idea has been pursued by the Russian school and appeared in a few isolated papers in the Western literature. See, for example, the 1971 article of

Tsung Na and Arthur Hansen[31].

Historically, the dimensional analysis originated with Lord Rayleigh who published an article on the "principles of similitude" and proposed a study of invariant properties of vibrations or of propagation of waves in his two volumes on "The theory of sound," which appeared in 1877 and 1878. There may be a counter-claim, crediting Fourier with this idea. Fourier offered early arguments advocating the use of physical dimensions in applied mathematics, but a systematic (deductive in the mathematical sense) approach to this subject should be credited to Lord Rayleigh in the opinion of this writer. Let me abstract an argument offered by Rayleigh.

Let us consider the differential equation of a vibrating mass m in the form

$$m\ddot{x} + kx = 0, \quad k \propto T/\ell$$

where m is mass concentrated in the center of a string of length ℓ, and the string has a constant tension T, then the period of one vibration τ is proportional to the quantity $\sqrt{\ell m/T} = \ell^{\frac{1}{2}} m^{\frac{1}{2}} T^{-\frac{1}{2}}$. Rayleigh concludes that even in non-linear vibrations, where all we can say about the period τ is that it must be a function of ℓ, m, T, we can conclude that $\tau \propto T^{-\frac{1}{2}} \ell^{\frac{1}{2}} m^{\frac{1}{2}}$.

Arguing this way Rayleigh introduces in his "Theory of Sound",Vol. II, other dimension less quantities, such as the Reynolds number given by $Re = vd/\nu$. (v is velocity, d principal dimension, ν-kinematic viscosity). Other dimensionless quantities bearing the names of Prandtl , Grasshoff, Fraude are associated with other important phenomena that are frequently encountered in the literature.

References for Chapter 4

[1] B. Noble, Complementary variational principles for boundary value problems, I, Basic Principles, Report #473, Math. Res. Center, University of Wisconsin, Madison, Wisc. (1964).

[2] B. Noble, Complementary variational principles II, Non-linear networks, Report #643, Math. Res. Center, Univ. of Wisconsin, Madison, Wisc., (1966).

[3] A.M. Arthurs, A note on Komkov's class of boundary value problems and associated variational principles, J. Math. Analysis Applications, 33, (1971), p.402-407.

[4] B. Noble and M.J. Sewell, On dual extremum principles in applied mathematics, J. Inst. Maths. Applics., 9, 1972, pp. 123-193.

[5] T. Kato, On some approximate methods concerning the operators T*T, Math. Ann, 126, (1953), p. 253-257.

[6] N. Bazley and D.W. Fox, Methods for lower bounds to frequencies of continuous elastic systems, John Hopkins Univ. Applied Physics Lab. Report TG-609, 1964.

[7] J.von Neumann, Über adjungierte functional Operatoren Ann., Math., 33, (1932), p. 294-310.

[8] B.N. Parlett, The symmetric eigenvalue problem, Prentice Hall, Englewood Cliffs, 1980.

[9] I. Herrera and M.J. Sewell, Dual extremal principles for non-negative, unsymmetric operators, J. Inst. Math. Applics. 21, (1978) p. 95-115.

[10] I. Herrera and J. Bielak, Dual variational principles for diffusion equations, Quart. Appl. Math., 85, 1976.

[11] W. Collins, Dual Extremum principles and Hilbert space decomposition in Duality in problems of mechanics of deformable bodies, Ossolineum, Warsaw, 1979.

[12] E. Tonti, On the variational formulation for linear initial value problems, Annali di Matematica pura ed applicata 95, p. 331-360, (1973).

[13] N. Dunford and J. Schwartz, Linear operators, Volume 1, Interscience, New York, 1962.

[14] S.L. Sobolĕv, Applications of Functional Analysis in Mathematical Physics, Amer. Math. Soc. Translation, Vol. 7, Rhode Island, (1963.(particularly pages 33-45).

[15] V. V. Novozhilov, Elasticity, Foundations of the non-linear theory of elasticity - Gostekhizdat, Moscow, 1948.

[16] N.I. Muskhelishvili, Some basic problems of the mathematical theory of elasticity, Noordhoff, Groningen, Holland, 1953.

[17] Goldstein H., Classical mechanics, Addison Wesley, Reading, Mass., 1950.

[18] S.Mikhlin, Linear equations of mathematical physics, Holt, Rinehart, and Winston, New York 2967.

[19] P.M. Morse, H. Feshbach, Methods of theoretical physics, McGraw-Hill, New York, 1953, Vol. I and II.

[20] H. Bateman, Notes on a differential equation which occurs in the two-dimensional motion of a compressible fluid and the associated variational problems, Proc. Roy. Soc., A125 (1929), 598.

[21] H. Bateman, Irrotational motion of a compressible inviscid fluid, Proc. Nat. Acad. Sci. USA, 16, (1930), p. 816-25.

[22] C.C. Lin and L.I. Rubinov, On the flow behind curved shocks, J. Math. Phys. 27(1948), p. 105-129.

[23] M. Shiffman, On the existence of subsonic flows of a compressible fluid, J. Rat. Mech. Anal. 1(1952), 605-616.

[24] P.E. Lush and T.M. Cherry, The variational method in hydrodynamics, Quarterly J. Anal. Mech. and Appl. Math. 9(1956), p. 6-21.

[25] A.S. Sumbatov, Non-extremality of a family of curves that are determined by the dynamic equations of a nonholonomic Chaplygin's system. The journal of differential equations (the Soviet journal, U.D.K.), Vol. II, #5, 1984, p. 897-899.

[26] R. Saeks, Resolution space, operators and systems, Lecture Notes in Economics and Mathematical Systems, Vol. 82, Springer-Verlag, Berlin, 1973.

[27] L.S. Pontrjagin, Topological Groups, Gordon Breach, New York, 1966.

[28] H. Hertz, Prinzipien der Mechanik in neuen Zusammenhang dargestellt, Barth Verlag, Leipzig, 1910.

[29] J.P. Aubin, Functional Analysis, Wiley, N.Y, 1979.

[30] R.A. Adams, Sobolëv Spaces, Academic Press, N.Y., 1975.

Chapter 5

Design Optimization

5.0 General remarks.

The problems of optimal design briefly discussed in the previous chapters have common mathematical foundations.

The problems of structural equilibrium or stability are generally modelled by elliptic partial differential equations. The cost functionals are either linear, quadratic (energy functionals) or neither, such as local functionals representing the maximum stress, the maximum deflection, or the magnitude of a buckling load. The constraints maybe either linear, or nonlinear functional inequalities. The linearity or non-linearity is with respect to the state of the system. The coefficients of the differential equations of the state depend on the design parameter vector $\underset{\sim}{u} \in U$, where U is the set of admissible designs. U is a convex, closed subset of a Sobolev space H. The functional analytic setting is as follows. Let $\Omega \subseteq \mathbf{R}^n$ be a bounded open region with Lyapunov boundary $\partial\Omega$. n = 1,2, or 3 represents the one, two or three-spatial dimensions of the domain Ω, respectively, that is occupied by the structure.

The state of the system is a vector in a Sobolev space $H = H_o^{\alpha}$ or $H = \overset{\circ}{W}{}_2^{\alpha}$ in the original Sobolëv notation. The inner product in H is denoted by $\langle\ ,\ \rangle_H$.(a small circle super or subscript denotes a compact support.)

A vector of design parameters (or "constants") $\underset{\sim}{u}$ is an element of a bounded set $U \subset W_p \subseteq L_2(\Omega)$, where $W_p(\Omega)$ is another Sobolëv space that is imbedded in $L_2(\Omega)$.

The set U obeys some constraints. It is called the set of admissible designs. Specifically U is bounded and is bounded away from zero. This implies that there exist positive numbers u_o and u_{max} such that for any $\tilde{u} \in U$ it is true that

$0 < u_o \leq || \tilde{u} || \leq u_{max}$.

In linear problems in which the state of the system is modelled by a linear differential equation of elliptic type, generally two basic bilinear energy forms determine the behavior of the system. Let us denote them by

$$\tilde{a}(w,v) \text{ and } \tilde{b}(w,v),$$

where w,v denote admissible states of the system, that is elements of H satisfying suitable boundary and/or initial conditions. We also insist that w,v are weak solutions of the state equation

$$(5.1) \quad L\, w(u) = q \quad ,$$

for some q, where $q \in Q \subset H^*$ (the topological dual of H, that is the space of continuous linear functionals mapping elements of H into $\mathbb{R}$) and Q is a bounded closed, convex subset of H^*. For example, if H is the Sobolev space $\dot{H}_o^2\ (\Omega)$ then H^* is $H^{-2}\ (\Omega)$.

Using Sobolëv imbedding lemma, we can claim that the imbedding of $H = W_2^{\alpha}\ (\Omega), \alpha > 1$ in $L_2(\Omega)$ is compact.

The basic design problems concern the choice of an admissible design $\hat{u} \in U$ such that the corresponding state $W(\hat{u})$ is a weak (or sometimes a classical) solution of the state equation (5.1), obeys all constraints assigned to the system, and assigns a minimum (or maximum) value to an assigned cost functional J(w).

The dynamic problems have an almost identical setting, but the "static" equation (5.1) is replaced by an eigenvalue problem

$$L\, w(u) = \lambda\, B\, w(u)\,, \qquad (5.2)$$

or

$$L\, w(u) - \lambda\, B\, w(u) = f, \qquad (5.2^a)$$

where B is a symmetric positive definite operator mapping H into H or H into H*.

In problems of elastic structures L is also self-adjoint.

(5.1) The bilinear energy forms.

The "variational form" of equations (5.1) and (5.2) is, respectively

$$(5.1^A) \quad < L\, w,\, v >_H \; = \; < q, v >_H \quad ,$$

and

$$(5.2^A) \quad < L\, w,\, v >_H \; = \lambda < B\, w,\, v >_H \quad .$$

The bilinear (or sequilinear) form $< L\, w, v >$ may be identified with the bilinear (or sequilinear) energy functional $\tilde{a}(\underset{\sim}{w},\underset{\sim}{v})$. If L is a self-adjoint operator then $\tilde{a}(\underset{\sim}{w},\underset{\sim}{v}) = \tilde{a}(\underset{\sim}{v},\underset{\sim}{w})$ (5.3^a)

Similarly, let $< B\, \underset{\sim}{w}, \underset{\sim}{v} >_H = b(\underset{\sim}{w},\underset{\sim}{v})$

$$= \tilde{b}\,(\underset{\sim}{v},\underset{\sim}{w}). \qquad (5.3^b)$$

We assume coercive property of the bilateral form $\tilde{a}(w,v)$.

Let us suppose the $\underset{\sim}{w}$ and $\underset{\sim}{v}$ are k-vectors, and that

$$\tilde{a}(\underset{\sim}{w},\underset{\sim}{v}) = \sum_{i,j=1}^{k} < c_{ij}\, L\, w_i,\, v_j >_H \, .$$

Then $\tilde{a}$ is $L_2\,(\Omega)$ coercive if

$$\tilde{a}(\underset{\sim}{w},\underset{\sim}{v}) = \sum_{i,j=1}^{k} c_{ij} < L\, w_i, v_j >_H =$$

$$\sum_{i,j} c_{ij} \, (P(L\, W_i),\, P(v_j))_{L_2} \,,$$

where for any k-vector $\underset{\sim}{\xi}$

$$\sum_{i,j=1}^{k} c_{ij}\, \xi_j \geq m \sum_{i=1}^{k} \xi_i^2 \quad \text{for some } m > 0, \quad \text{and where}$$

P is the operator of embedding of H in L_2. The scalar case is taken care of by assuming that k = 1. The theory of design optimization for coercive forms is very attractive. Solutions exist, and iterative schemes converge, provided the boundary and/or initial value problems are well posed in the sense of Hadamard.

For example, let $\underset{\sim}{w} \in H$, $\underset{\sim}{u} \in U$, $\underset{\sim}{f} \in H^*$, where $H = \overset{o}{W}{}_2^{\ell}$. The state equation is given in the variational form

$$\tilde{a}(\underset{\sim}{w}(u),\, \underset{\sim}{v}) = (\underset{\sim}{f}, \underset{\sim}{v}), \qquad (5.4)$$

where $\underset{\sim}{v} \in H$.

Lemma 5.1

Let $\tilde{a}$ be coercive, bilinear symmetric form on H and $\underset{\sim}{w}(\underset{\sim}{u})$ be a linear function of $\underset{\sim}{u}$. Let $\underset{\sim}{u}$ be any admissible design (this means that $\underset{\sim}{w}(\underset{\sim}{u})$ is an admissible displacement, or admissible state of the system) for any well-posed boundary value problem. Then the mapping $\underset{\sim}{u} \to \underset{\sim}{w}(\underset{\sim}{u})$ is one to one. The Lax-Milgram theorem can be invoked for the bilinear functional $\tilde{a}(\underset{\sim}{w},\underset{\sim}{v}) = A(\underset{\sim}{u},\underset{\sim}{v})$. A difficult and frequently quite subtle problem concerns continuous dependence of the state on the admissible design.

If the inhomogeneous term $f(\underset{\sim}{x})$ is independent of the design $\underset{\sim}{u} \in U$, then the nature of the operator $L(\underset{\sim}{u})$ completely determines the dependence of the state $\underset{\sim}{w}(\underset{\sim}{u})$ on $\underset{\sim}{u} \in U$.

Special case includes problems of beam, plate

and shell theory when the forms $\breve{a}(w,v)$, $\tilde{b}(w,v)$ depend explicity on h and may be written as

$$\tilde{a}(w,v) = \int_\Omega \phi_1(\underset{\sim}{u}) \sum_{i,j=1}^{k} a_{ij}(P_i u, P_j v)$$

$$+ \int_\Omega \phi_2 \ (u) \sum_{i,j=1}^{k} b_{ij} \ (P_i u, \ P_j v)$$

$$a_{ij} = a_{ji}, \ b_{ij} = b_{ji} \in L_\infty(\Omega),$$

and ϕ_i, $i = 1,2$ are continuous.

This case was discussed by V.G. Litvinov in [1]. Then the coercivity assumptions

$$\sum_{i,j=1}^{k} a_{ij} \ \xi_i \ \xi_j \geq c \sum_{i=1}^{c} \xi_i^2$$

$$\sum_{i,j} b_{ij} \ \xi_i \ \xi_j \geq c \sum_i \xi_i^2 \ , \ \text{for some } c > 0$$

imply that $a(w,w) \geq c \ ||w||_H^2$, $w \in H$

for all $u \in U$,

and weak convergence $u_i \to \tilde{u}$ in B implies weak convergence of $w_i \to \tilde{w}(\tilde{u})$ in H.

5.2 The choice of Hilbert spaces .

5.2.1 Linear elasticity.

In Chapter 2 of volume 1, pages 156-159, we discussed the Legendre transformation relating the equations of equilibrium to the equations of equilibrium to the equations of compatibility. If rotation compenents are ignored (supposedly, they are "small") and only linear strains are considered $e_{ij} = \frac{1}{2} (w_{i,j} + w_{j,i})$, $i = 1,2,3$, $i \neq j$,

$e_{ii} = e_i = w_{i,i}$, where $\underset{\sim}{w} = \{w_1, w_2, w_3\}$ is the displacement vector.

The simplest versions of the equilibrium and compatilbility equations are

$$\left.\begin{aligned} &(5.5^a) \quad A^*\underset{\approx}{\tau} = \tau_{ij,j} = -\rho_i \\ &(5.5^b) \quad A\underset{\sim}{w} = \underset{\approx}{\varepsilon} \end{aligned}\right\} \text{ in } \Omega \subseteq R^3.$$

The constitutive equations are given by Hooke's law

$$(5.5^c) \quad \underset{\approx}{\tau} = \underset{\approx}{C} \cdot \underset{\approx}{\varepsilon} = \underset{\approx}{C} \cdot A\,\underset{\sim}{w}, \quad \text{in } \Omega.$$

For the sake of convenience we can regard τ and ε as six-dimensional vectors, rather than tensors of rank two.

Conditions on the boundary $\partial\Omega$ of the region Ω can be of the first kind, according to Muskhelishvili's classification, i.e.

$$(B) \quad \tau_{ij} \cdot n_j = \hat{\tau}_i \text{ on } \partial\Omega.$$

Here the design variables could be the components of $\underset{\approx}{C}$ in (5.5^c). In that case "design" consists of choosing suitable materials to use in a multi-layer, laminated or inhomogeneous medium when we optimize some cost functional, such as the maximum stress.

5.2.2 The Form A*A

The use of duality is particularly effective in problems involving design optimization of the state equation is of the form

$$(5.6) \quad L(\underset{\sim}{u})\, w(\underset{\sim}{x}) = f(\underset{\sim}{x}, \underset{\sim}{u})$$

where L is of the form

$$(5.7) \quad L = A^*A, \text{ (or } A\,A^*,$$

and $u(\underset{\sim}{x})$ is a design vector.

The problems of classical elasticity, beam or plate theory are most naturally formulated in the $H^2(\Omega)$ Sobolev space setting.

Then L maps $H^2(\Omega)$ into $H_0^{-2}(\Omega)$ which is the topological dual of $H^2(\Omega)$.

Physically w(x) is the generalized displacement, $f(\underset{\sim}{x},\underset{\sim}{u})$ the corresponding generalized force.

The bilinear energy form is

(5.8) $\tilde{a}(w,\psi) = \langle L\,w,\ \psi\rangle$,

since $L = A^*A$

(5.9) $\tilde{a}(w,\psi) = \langle\ Lw,\psi\rangle = \langle\ Aw,\ A\psi\rangle$.

A few remarks concerning operators of continuum mechanics should be made first. Partially these remarks duplicate a section of volume I of this work.

5.3 Examples of functional-analytic duality in structural problems.

Practical design of vibrating structures has been carried out on trial and error basis. No comprehensive theory exists at the present time, although some progress in optimizing the design of single elements has been made. Prager and his collaborators worked primarily on optimization of the design of a single beam, making use of the Betti-Castigliano deflection formula and deriving some theoretical results for optimization of weight for a given deflection at a known point along the length of the beam, and related problems (see [2] or [3]). A more difficult problem of simultaneous optimal control was considered by Komkov and Coleman in [4], however the applications of their work were limited, and no general approach theory was derived.

A different dynamic optimization was pursued by Haug [5], Haug et al [6], and others, who attempted to derive algorithms for gradual improvement in design by an iterative process. So far the numerical results obtained at the University of Iowa are promising, and have been applied successfully to some problems of weapon design. The purpose of this work is to derive results analogous to [6] for the dynamic case, using functional analytic arguments outlined in the original work of Vainberg [7], and related to the theory of optimization in normed spaces. (See Nashed [8], Kato [9], or the original papers of Fréchet [10] and Gateaux [11]). While the mathematical theory utilized in this section is known, and no new mathematical results are derived, to the best of our knowledge this is one of the first direct applications of this body of mathematical knowledge to actual problems of structural dynamics. Morever this is not a straight-forward application of the abstract differentiation to structural problems. The emphasis lies in dependence of operators occurring in the equations of motion on the eigenvalue ξ, and on determining the sensitivity of the system with respect to changes in ξ.

Using Fréchet differentiation the sensitivity of beams and plates is established (to changes in design). Extension of these results to generalized structures is theoretically straight-forward, but computationally complex. Some computational aspects are discussed here. Optimization is carried out on fairly simple examples using a version of the steepest descent algorithm. However, the basic approach to to optimization theory carried out here consists of deriving the necessary condition of optimiality by direct Fréchet differentiation. A particular application of this technique to a design of a gun tube was given. The theoretical justification, and the rules for formal Fréchet differentiation and explanations concerning properties of higher order Fréchet derivatives are given in Volume 1 of this work.

5.4. A class of optimal design problems described by a distributed parameter system.

Consider a design problem postulated on a subset $\Omega \subset R^k$, with a local coordinate system $\underset{\sim}{x} = (x_1, x_2, \ldots, x_k)$. The state of the system is determined by an n-dimensional vector function $\underset{\sim}{z}(\underset{\sim}{x})$, which is an element of a Hilbert space H_1.

The design of the system is described by an m-dimensional vector function $\underset{\sim}{u}(\underset{\sim}{x})$. There are constraints on stress, deflection, etc., and physical constrains on the design vector $\underset{\sim}{u}(\underset{\sim}{x})$. The set of m-dimensional vector functions satisfying the constaints will be called the set of admissible designs and will be denoted by $U(\underset{\sim}{x})$.

The state vector $\underset{\sim}{z}$ satisfies a system of differential equations

$$L(\underset{\sim}{u})\underset{\sim}{z} = Q(\underset{\sim}{x}, \underset{\sim}{u}), \quad \underset{\sim}{x} \in \Omega, \tag{5.1}$$

where the forcing function Q is an element of a Hilbert space H_3. Thus $L: H_1 \to H_3$. The state variable $\underset{\sim}{z}$ also satisfies a set of boundary conditions

$$B\, \underset{\sim}{z}(\underset{\sim}{x}) = \underset{\sim}{q}(\underset{\sim}{x}), \quad \underset{\sim}{x} \in \partial\Omega, \tag{5.2}$$

We introduce also Hilbert spaces H_2, H_4 of functions whose domain is $\partial\Omega$, and such that the boundary condition (5.2) is given as a mapping $B: H_2 \to H_4$.

While the same symbol z is used in relations (5.1) and (5.2) it actually depicts different classes of functions. However, $\underset{\sim}{z}$ regarded as an element of $H_1 \oplus H_2$ is continuous on $\Omega \cup \partial\Omega$ and no problems will arise due to the simplified notation.

The system (5.1) and (5.2) will be referred as

B.V.P. (boundary-value problem).

Let A denote the operator

$$A = \begin{bmatrix} L & 0 \\ 0 & B \end{bmatrix} : H_1 \oplus H_2 \to H_3 \oplus H_4.$$

(i.e. A is a 2x2 matrix of operators).

The existence of a unique solution of B.V.P. is equivalent to the statement that A^{-1} is defined. As was already indicated, well posedness of the problem, in the sense of Hadamard, is equivalent to the statement that A^{-1} is defined and is a bounded operator from $H_3 \oplus H_4$ into $H_1 \oplus H_2$. This

is a well known property of stable structural designs. In an important class of structural designs, the differential operator L also turns out to be positive definite and bounded below (away from zero), and L^{-1} is a completely continuous operator. A^{-1} is generally given by a formula of the type (5.1^a)

$$\underset{\sim}{Z}(x) = \int_{\Omega} K(x,X)\, L^*(Z(X))\, d\underset{\sim}{X} + \int_{\partial\Omega} B(K(x,X))\, Z(X)\, d\underset{\sim}{s},$$

the Kernel K(x,X) uniquely determining the form of A^{-1}.

Also of interest in this work are designs of structural members subjected to buckling and to natural frequency constraints.

In each of these cases, the physical problem leads to an eigenvalue system

$$L'(\underset{\sim}{u})\, \underset{\sim}{y}(\underset{\sim}{x}) = \zeta\, M(\underset{\sim}{u})\, \underset{\sim}{y}(\underset{\sim}{x}), \quad \underset{\sim}{x} \in \Omega, \qquad (5.3)$$

$$B'\, \underset{\sim}{y}(\underset{\sim}{x}) = 0 \qquad , \quad \underset{\sim}{x} \in \partial\Omega, \qquad (5.4)$$

where L' and M are symmetric, positive, bounded below differential operators, ζ is a real eigenvalue, and y(x) is the corresponding eigenfunction. The eigenfunctions $\underset{\sim}{y}(\underset{\sim}{x})$ are normalized by the condition

$$\langle M\underset{\sim}{y},\underset{\sim}{y}\rangle = |||\underset{\sim}{y}|||^2 = 1,$$

where $|||\cdot|||$ denotes the energy norm induced by the inner product $[\underset{\sim}{X},\underset{\sim}{Y}] = \langle MX,Y\rangle$. The inner product $\langle \underset{\sim}{X},\underset{\sim}{Y}\rangle$ is defined in the usual way, i.e.

$$\langle \underset{\sim}{X},\underset{\sim}{Y}\rangle = \int_\Omega \underset{\sim}{X}(\underset{\sim}{x})\ \underset{\sim}{Y}(\underset{\sim}{x})\ d\underset{\sim}{x}.$$

One wishes to minimize, or approximately minimize, a functional

$$J : \{H_1 \oplus H_2 \times U\} \to \mathbb{R}.$$

where J is usually either an L_2 norm of Y, an energy norm, or a possibly non-linear, but positive functional. Analytically, J is of the form

$$J(\zeta, x_z(\underset{\sim}{u}), \underset{\sim}{u}) = h_0(\zeta) + \int_{\partial\Omega} g_0(\underset{\sim}{z},\underset{\sim}{x})\ d\underset{\sim}{x} + \int_\Omega f_0(\underset{\sim}{z}(\underset{\sim}{u}),\ \underset{\sim}{u},\underset{\sim}{x})\ d\underset{\sim}{x}. \tag{5.5}$$

The design problem may thus be reduced to selection of an admissible design $\underset{\sim}{u}(\underset{\sim}{x}) \in U$ to minimize (or approximately minimize) J, subject to the constraints such as (B) and performance constraints of the form

$$\psi_\alpha = h_\alpha(\zeta) + \int_{\partial\Omega} g_\alpha(\underset{\sim}{z},\underset{\sim}{x},\underset{\sim}{u})\,d\underset{\sim}{x} + \int_\Omega f_\alpha(\underset{\sim}{z},\underset{\sim}{x},\underset{\sim}{u})\,d\underset{\sim}{x} \begin{cases} =0, & \alpha=1,2,\dots r^1, \quad (5.6^a) \\ \le 0, & \alpha=r^1+1,\dots,r. \quad (5.6^b) \end{cases}$$

$$\phi_\beta(\underset{\sim}{x},\underset{\sim}{u}) \begin{cases} =0, & \beta=1,2,\ldots\xi^1, \\ \leq 0, & \beta=\xi^1+1,\ldots,\xi. \end{cases} \quad \left\{ \underset{\sim}{x} \in \Omega \right. \qquad \begin{matrix} (5.7^a) \\ (5.7^b) \end{matrix}$$

Comments based on computing experience.

The pointwise constraints ($5.7^{\underline{a}}$) and ($5.7^{\underline{b}}$) are allowed to depend only on the design variable u. If the state variables z were present, computational difficulties would arise later in the development.

Constraints of the form $\eta(\underset{\sim}{x},\underset{\sim}{u},\underset{\sim}{z}) \leq 0$, $\underset{\sim}{x} \in \Omega$, arise often, however, and must be treated. It is readily seen that the pointwise constraint $\eta(\underset{\sim}{x}) \leq 0$ is equivalent to the functional constraint

$$\int_\Omega (\eta(\underset{\sim}{x}) + |\eta(\underset{\sim}{x})|)\, d\underset{\sim}{x} = 0. \qquad (5.8)$$

Similarly, $\phi(\underset{\sim}{x}) \leq a_0 \ \forall \underset{\sim}{x} \in \Omega$ is equivalent to

$$\int_\Omega \{(\phi(\underset{\sim}{x}) - a_0) + |\phi(\underset{\sim}{x}) - a_0|\}\, d\underset{\sim}{x} = 0.$$

5.5 Computation of Gateaux derivatives associated with an optimal design problem. (an informal discussion)

The standard definition (see [11]) of a Gateaux derivative of a function $f\colon B_1 \to B_2$ in the direction of a vector h is employed. Here B_1, B_2 are Banach spaces, $h \in B_1$. If for a sufficiently small value of the real parameter t

$$f(x_0+th) - f(x_0) = t\, L_{x_0} + r(x_0,th), \qquad (5.9)$$

where

$$\lim_{t \to 0} \left\| \frac{r(x_0, th)}{t} \right\| = 0 \quad , \qquad (9\underline{a})$$

and where L_{x_0} is a linear operator; $L_{x_0} : B_1 \to B_2$;

we assume that f is Gateaux differentiable at x_0 (ϵB_1), and the operator L_{x_0} (the Gateaux derivative of f in the direction of h) is defined.

Here B_1 and B_2 are Hilbert spaces and, in the development that follows, B_1 will be identified with the Sobolev space H^k, while in most cases, B_2 will be the real line.

The minimization problem consists of finding a minimum of a functional J which in many problems of structural analysis is a quadratic functional, (L_2 norm of the state variable z, an energy norm, or possibly other norm) subject to constraints imposed on the state variable z, which must satisfy the appropriate differential equations of mathematical physics. That is, the problem is formulated as a minimization problem for the functional $J(u,z,\xi,x)$ subject to differential equations $Lz = \xi z$. For example,

$Lz = \xi z$ may be of the form $\frac{d^2}{dx^2}[EI(x) \frac{d^2 z}{dx^2}] + \zeta z = 0.$

Then the minimization problem may be restated in an enlarged space as that of finding the minimum of

$$\lambda_1 \, [\langle EI(x) \frac{d^2 z}{dx^2} , \frac{d^2 z}{dx^2}\rangle + \langle \zeta z, z\rangle] + \lambda_0 J,$$

where λ_0, λ_1 are Lagrangian multipliers.

Hence the constraint equation is regarded as one of the Euler-Lagrange equations, which (subject to some smoothness assumptions on the state variable z) is a necessary condition for the solution of the minimum problem in the enlarged space.

Consider the functional $J(\tilde{u},\tilde{z},\tilde{x},\tilde{\zeta})$ defined by (5.5) and its Gateaux variation δJ, where $\delta h = th$ (for sufficiently small (t) and $\delta h = (\delta z, \delta u, \delta\zeta)$. In a purely formal fashion one may write

$$\delta J = \sum_{\alpha} \frac{\partial h_\alpha}{\partial \zeta}\delta\zeta + \left\langle \frac{\partial g_\alpha}{\partial \tilde{z}}, \delta\tilde{z}\right\rangle_{\partial\Omega} + \left\langle \frac{\partial f_\alpha}{\partial \tilde{z}}, \delta\tilde{z}\right\rangle_{\Omega} + \left\langle \frac{\partial f_\alpha}{\partial \tilde{u}}, \delta\tilde{u}\right\rangle_{\Omega} + r(\tilde{h}), \tag{5.10}$$

where $\langle\cdot,\cdot\rangle_{\partial\Omega}$ and $\langle\cdot,\cdot\rangle_{\Omega}$ are the appropriate inner products of functions whose domains are $\partial\Omega$ and Ω, respectively, and as usual

$$\lim_{\|\tilde{h}\|\to 0} \frac{\|r(\tilde{h})\|}{\|\tilde{h}\|} = 0.$$

The problem arises of justifying this formal approach by showing differentiability of the operator L' with respect to the eigenvalue ζ. There are some obvious difficulties, since L' is an unbounded operator. Instead, the fact that A^{-1}, defined by (5.1^a) is a compact operator in the common problems of structural analysis, can be employed. Hence, the eigenvalues of A^{-1}, which are ζ^{-1}, are a discrete subset of the real line, i.e. the spectrum of A^{-1} consists only of the isolated eigenvalues. One may now quote the theorem on bounded invertibility (see Kato [9] pp. 196-197) and a well known result on perturbation of spectrum of A^{-1}; namely that the spectrum of A^{-1} changes continuously with A^{-1}, such small changes being interpreted in the usual operator norm topology([9]).

For example, consider a beam equation of the form

$$\left[\frac{d^2}{dx^2}\sqrt{EI}\left(\sqrt{EI}\frac{d^2}{dx^2}\right)\right] y + q_n(x)y = f(x) + \xi_n y \quad (5.11)$$

subject to appropriate boundary conditions.

Let $q_n \xrightarrow{L_2[0,1]} q$, $f \in L_2[0,1]$. Then $\mathfrak{R}(\xi_n) \to \mathfrak{R}(\xi)$, where $\mathfrak{R}$ is the resolvent, and $\lim_{n\to\infty} \xi_n = \xi$, and ξ does not lie in the convex hull of ξ_n implies that the solutions of (11) converge in $L_2[0,1]$ to the solution of the equation

$$\left[\frac{d^2}{dx^2}\sqrt{EI}\left(\sqrt{EI}\frac{d^2}{dx^2}\right)\right] y + q(x)y = f(x) + \xi y. \quad (5.12)$$

The key in this discussion is the avoidance of singular perturbations.

To illustrate this remark consider the following two examples. The fourth order beam operator

$$\frac{d^2}{dx^2}\left(EI(x)\frac{d^2}{dx^2}\right), \quad (5.13)$$

with boundary conditions of either free support at both ends, or cantelevered at one end, implies continuous changes in the eigenvalues with a continuous change in EI(x). The operator

$$\left(\frac{d^2}{dx^2} EI(x) \frac{d^2}{dx^2} - \alpha^2\frac{d^2}{dx^2}\right) \quad (5.14)$$

$$\text{with} \quad \begin{cases} y(0) = y'(0) = 0 \\ y(1) = y'(1) = 0, \end{cases} \qquad (5.14^a)$$

on the other hand, will cause singular perturbations.
(Note: This is exactly the example of singular perturbation given in Kato's book [9], page 436, example 1.20).

That is, the eigenvalues of (5.13) are stable and of (5.14) are not.
Note: The discussion of stability of eigenvalues can be traced to Lord Rayleigh (see [25], page 300).

It is assumed in the use of formula (5.12) that A^{-1} is compact and that perturbations of eigenvalues of A^{-1} (and of A) are stable. That is, one is dealing with a common case of stable structural design.

5.6 Manipulation of formula (5.10).

The meaning of $\delta\zeta$ and its relationship to δu needs to be determined next in the interpretation of formula (5.10). Regarding $\delta\zeta$ as the Gateaux variation of the first eigenvalue of the system $L'y = \zeta My$, and assuming that L' and M depend holomorphically on $\tilde{u}$, it is possible to apply Aronszajn's theory of general quadratic forms (see [12]) and justify formal differentiation of the associated Rayleigh quotient. (see Kato [9], VII - 83, pp. 421-422).

The first eigenvalue of the system is equal to the value of the Rayleigh quotient

$$\zeta_1 = \frac{\langle L'\tilde{y}_1, \tilde{y}_1\rangle_\Omega}{\langle M\tilde{y}_1, \tilde{y}_1\rangle_\Omega},$$

where $y_1(x)$ is the corresponding eigenfunction.

Assuming symmetry of L' and M and positive definiteness of M, and using formal rules of Gateaux or Fréchet differentiation (see for example the introductory notes in [13]) for the Rayleigh functional, one obtains

$$\zeta_{1_h} = \langle M\underset{\sim}{y}_1, \underset{\sim}{y}_1\rangle_\Omega^{-1} \{\langle (L'\underset{\sim}{y}_1)_h, \underset{\sim}{y}_1\rangle_\Omega + \langle L'\underset{\sim}{y}_1, \underset{\sim}{y}_{1_h}\rangle_\Omega - \langle L'\underset{\sim}{y}_1, \underset{\sim}{y}_1\rangle$$

$$(\langle M\underset{\sim}{y}_1, \underset{\sim}{y}_1\rangle_\Omega)^{-1} [\langle (M\underset{\sim}{y}_1)_h, \underset{\sim}{y}_1\rangle_\Omega + \langle M\underset{\sim}{y}_1, \underset{\sim}{y}_{1_h}\rangle_\Omega]\}$$

$$= \langle M\underset{\sim}{y}_1, \underset{\sim}{y}_1\rangle_\Omega^{-1} \{\langle (L'\underset{\sim}{y}_1)_h, \underset{\sim}{y}_1\rangle - \zeta\langle (M\underset{\sim}{y}_1)_h, \underset{\sim}{y}_1\rangle + \langle (L'\underset{\sim}{y}_1 -$$

$$\zeta M\underset{\sim}{y}_1), \underset{\sim}{y}_{1_h}\rangle\}, \qquad (5.15)$$

where subscript h denotes Fréchet differentiation with respect to a typical component h of the vector u. The last two terms in (5.15) may be omitted since their sum is zero.

Expressions of the type $(L'\underset{\sim}{y})_h$ have to be carefully interpreted, since $\underset{\sim}{y}_1$ and the map L' both depend on h. If both are Fréchet differentiable, the following rule is easily justified

$$\delta(L\underset{\sim}{y}) = (\delta L)\underset{\sim}{y} + L\delta\underset{\sim}{y} \qquad ((5.16)$$

or equivalently

$$(L\underset{\sim}{y})_h = L_h\underset{\sim}{y} + L\underset{\sim}{y}_h$$

where $L_h y$ has to be interpreted as the Gateaux variation of the operator L with y regarded as fixed, while y_h is the corresponding variation of y.

Observe: $L(h_0+th)\underset{\sim}{y}(h_0+th) - L(h_0)\underset{\sim}{y}(h_0)$

$$= [L(h_0+th)\underset{\sim}{y}(h_0+th) - L(h_0)\underset{\sim}{y}(h_0+th)] + [L(h_0)\underset{\sim}{y}(h_0+th) - L(h_0)\underset{\sim}{y}(h_0)].$$

The Gateaux variation of the first eigenvalue $\delta\zeta_1$ is given by

$$\delta\zeta_1(u) = \left\langle M\underset{\sim}{y}_1, \underset{\sim}{y}_1\right\rangle_\Omega^{-1} \left\{\left\langle \frac{\partial(L'\underset{\sim}{y}_1)}{\partial\underset{\sim}{u}}, \underset{\sim}{y}_1\right\rangle_\Omega - \zeta\left\langle \frac{\partial(M\underset{\sim}{y}_1)}{\partial\underset{\sim}{u}}, \underset{\sim}{y}_1\right\rangle_\Omega \delta\underset{\sim}{u}\right.$$

$$\left. + \left\langle (L'\underset{\sim}{y}_1 - \zeta M\underset{\sim}{y}_1), \frac{\partial\underset{\sim}{y}_1}{\partial u}\right\rangle_\Omega \right\} \delta\underset{\sim}{u}.$$

Since $\underset{\sim}{y}_1$ is the eigenfunction of (5.3) the last term can be omitted, and

$$\delta\zeta_1 = \left\langle M\underset{\sim}{y}_1, \underset{\sim}{y}_1\right\rangle^{-1} \left\{\left\langle \frac{\partial(L'\underset{\sim}{y}_1)}{\partial u}, \underset{\sim}{y}_1\right\rangle_\Omega - \zeta\left\langle \frac{\partial(M\underset{\sim}{y}_1)}{\partial u}, \underset{\sim}{y}_1\right\rangle\right\} \delta\underset{\sim}{u}. \tag{5.17}$$

<u>Note</u>: $\left\langle \frac{\partial(L'\underset{\sim}{y}_1)}{\partial u}, \underset{\sim}{y}_1\right\rangle_\Omega$ is not a scalar, it is a vector in the space dual to $\underset{\sim}{u}$. The same remark applies to the vector $\left\langle \frac{\partial M\underset{\sim}{y}_1}{\partial \underset{\sim}{u}}, \underset{\sim}{y}_1\right\rangle_\Omega$. Observe that $\left(\frac{\partial(My_1)}{\partial\underset{\sim}{u}}\right)$, $\left(\frac{\partial(L'y_1)}{\partial\underset{\sim}{u}}\right)$ are tensor products in $B_2^* \times U^*$ where $Ly_1 \in B_2$, $\underset{\sim}{u} \in U$ and B_2^*, U^* are spaces dual respectively to B_2 and U.

5.7. <u>An adjoint problem.</u>

Since the generalized displacement (or state) vector $\underset{\sim}{z}$ is an element of the Hilbert space $H_1 \oplus H_2$, its dual (the generalized force) is an element of the same Hilbert space. The duality is maintained only for the sake of physical interpretation.

Let $\underset{\sim}{\lambda}$ be an element of the dual space, satisfying the differential equation

$$L^*\underset{\sim}{\lambda}^J = \frac{\partial f_\alpha}{\partial \underset{\sim}{z}} \quad \text{in } \Omega , \tag{5.18}$$

where L* is a formal adjoint of L, satisfying

$$\left\langle \underset{\sim}{W}, L\underset{\sim}{z} \right\rangle_\Omega = \left\langle L^*\underset{\sim}{W}, \underset{\sim}{z} \right\rangle_\Omega + \left\langle C\underset{\sim}{W}, \underset{\sim}{z} \right\rangle_{\partial\Omega} , \tag{5.19}$$

where C is a linear operator, C: $H_2 \to H_2$. $\underset{\sim}{\lambda}^J$ satisfies boundary condition

$$\left\langle (C\underset{\sim}{\lambda}^J - \frac{\partial g_\alpha}{\partial \underset{\sim}{z}}), \delta\underset{\sim}{z} \right\rangle_{\partial\Omega} = 0, \tag{5.20}$$

for every admissible $\delta\underset{\sim}{z}$, which is certainly satisfied if

$$C\underset{\sim}{\lambda}^J = \frac{\partial g_\alpha}{\partial \underset{\sim}{z}} \quad \text{on } \partial\Omega . \tag{5.21}$$

Hence,

$$\left\langle L\delta\underset{\sim}{z}, \underset{\sim}{\lambda}^J \right\rangle_\Omega = \left\langle \frac{\partial f_\alpha}{\partial \underset{\sim}{z}}, \delta\underset{\sim}{z} \right\rangle_\Omega + \left\langle C\underset{\sim}{\lambda}^J, \delta\underset{\sim}{z} \right\rangle_{\partial\Omega}$$

and

$$\delta J = \frac{\partial h_\alpha}{\partial \zeta}\delta\zeta + \left\langle \frac{\partial g_\alpha}{\partial \underset{\sim}{z}}, \delta\underset{\sim}{z} \right\rangle_{\partial\Omega} + \left\langle \frac{\partial f_\alpha}{\partial \underset{\sim}{z}}, \delta\underset{\sim}{z} \right\rangle_\Omega + \left\langle \frac{\partial f_\alpha}{\partial \underset{\sim}{u}}, \delta\underset{\sim}{u} \right\rangle_\Omega$$

$$= \frac{\partial h_\alpha}{\partial \zeta}\delta\zeta + \left\langle \frac{\partial f_\alpha}{\partial \underset{\sim}{u}}, \delta\underset{\sim}{u} \right\rangle_\Omega + \left\langle \underset{\sim}{\lambda}^J L\,\delta z + \frac{\partial (Lz)}{\partial u}\delta u - \frac{\partial Q}{\partial u}\delta u \right\rangle_\Omega$$

$$+ \left\langle \frac{\partial g_\alpha}{\partial \underset{\sim}{z}}, \delta\underset{\sim}{z} \right\rangle_{\partial\Omega} - \left\langle C\underset{\sim}{\lambda}^J, \delta\underset{\sim}{z} \right\rangle_{\partial\Omega}$$

$$= \frac{\partial h_\alpha}{\partial \zeta}\,\delta\zeta + \left\langle \left(\frac{\partial f_\alpha}{\partial \underset{\sim}{u}} + \underset{\sim}{\lambda}^J \left(\frac{\partial (L\underset{\sim}{z})}{\partial u} - \frac{\partial Q}{\partial \underset{\sim}{u}}\right)\right), \delta\underset{\sim}{u} \right\rangle_\Omega$$

$$= \frac{\partial h_\alpha}{\partial \zeta}\,\delta\zeta + \left\langle \bar{\underset{\sim}{\Lambda}}, \delta\underset{\sim}{u} \right\rangle_\Omega . \qquad (5.23)$$

$$\bar{\underset{\sim}{\Lambda}} = \frac{\partial f_\alpha}{\partial \underset{\sim}{u}} + \underset{\sim}{\lambda}^J \left(\frac{\partial (L\underset{\sim}{z})}{\partial \underset{\sim}{u}} - \frac{\partial Q}{\partial \underset{\sim}{u}}\right) \qquad (5.24)$$

is the sensitivity vector. Suppose that $\delta\zeta \approx 0$, and

$\frac{\delta h_\alpha}{\delta \zeta}\,\delta\zeta \approx 0$, i.e. the first natural frequency

does not change significantly nor do admissible

changes in the design. Then $\delta J \simeq \langle \bar{\Lambda}, \delta u \rangle_\Omega$.

If the term $\left(\frac{\partial h_\alpha}{\partial \zeta} \cdot \delta\zeta\right)$ can not be ignored, a

substitution of (5.17) into (5.23) yields

$$\delta J = \frac{\partial h_\alpha}{\partial \zeta} \left\langle M\underset{\sim}{Y}_1, \underset{\sim}{Y}_1 \right\rangle^{-1} \left\{ \left\langle \left[\frac{\partial (L'\underset{\sim}{Y}_1)}{\partial \underset{\sim}{u}} \underset{\sim}{Y}_1 - \zeta \frac{\partial (M\underset{\sim}{Y}_1)}{\partial \underset{\sim}{u}} \underset{\sim}{Y}_1\right], \delta\underset{\sim}{u} \right\rangle_\Omega \right\}$$

$$+ \left\langle \bar{\underset{\sim}{\Lambda}}, \delta\underset{\sim}{u} \right\rangle_\Omega = \left\langle \underset{\sim}{\Lambda}^J, \delta\underset{\sim}{u} \right\rangle_\Omega . \qquad (5.25)$$

$\underset{\sim}{\Lambda}^J$ is the sensitivity vector for optimization of the cost functional J.

It is significant that no boundary terms (i.e. products in H_2) occur in the formula (5.25).

5.8 <u>An example</u>. Consider the dynamic problem of thin plate theory represented by the equation

$$\nabla^2 (D(x,y)\nabla^2 W) + \rho \frac{\partial^2 W}{dt^2} = Q(x,y)e^{j\omega t}, \ \underset{\sim}{x} \in \Omega, \ t \geq 0.$$

After separation of variables, this is reduced to an eigenvalue problem

$$(L' + \zeta I)\, W(\tilde{x}) = Q(\tilde{x}) \qquad \tilde{x} = \begin{bmatrix} x \\ y \end{bmatrix}. \tag{5.26}$$

where $L' = \nabla^2 (D(x,y)\nabla^2)$ and $M \equiv I$ is the identity operator.

Design a plate of a given weight (say unity) such that the natural frequency ζ takes an approximately extremal value. The design parameter $u(\tilde{x})$ is the thickness of the plate. The plate is clamped on the boundary. In fact this problem is equivalent to looking for a low approximate value of the sensitivity vector $\tilde{\Lambda}$.

We minimize

$$J = \mu_o \int_\Omega u(\tilde{x})\, d\tilde{x} + k\zeta \tag{5.27}$$

where μ_0 is a Lagrangian multiplier, and where k is a known constant. (Recall that ζ is proportional to ω^2). Identifying $h_0(\zeta)$ as $k\zeta$, $g_0 \equiv 0$, $f_0 = \mu_o u$, the sensitivity vector is computed as

$$\Lambda^J = \mu_o + k \langle y_1, y_1\rangle^{-1} \{\langle \nabla^2 (\frac{Eu_0^2}{4(1-\nu^2)} \nabla^2 y_1 + \frac{Eu_0^3}{12(1-\nu^2)} \frac{\partial(\nabla^2 y_1)}{\partial u}), y_1\rangle - \zeta \langle \frac{\partial y_1}{\partial u}, y_1\rangle\}. \tag{5.28}$$

Supposing that ζ_1 and $y_1(x)$ are known and $\frac{\partial y_1}{\partial \tilde{u}}$ is computed, the sensitivity vector (5.28) can be computed in terms of known physical quantities. An optimum design requires near zero sensitivity. In order that u_0 is optimal it is necessary for Λ^J to vanish. However equation (5.28) is difficult to solve for u_0 and gradual improvement approach is generally used for sub-optimal design. If Λ^J is not zero then following a steepest descent approach one could choose δu in the direction of $-\Lambda^J$. This is locally optimal direction. In general, the constraints imposed by physical consideration may not be satisfied by altered design vector.

Various versions of the gradient-projection technique have been used to correct this defect.

For an arbitrarily chosen design, the decrease in the magnitude of the sensitivity vector indicates the degree of improvement in the corresponding step of the steepest descent algorithm.

5.9 Shape Discontinuities

Shape optimization is a particularly difficult and treacherous problem. The very first serious mathematical discussion of the problem by Lagrange (reference [14]) contained some serious errors and even the theories developed for the simple one-dimensional problem of column optimiation had several flawed results published by "big name" researchers.

An ingenious approach of Hadamard [15] using properties of the Green's function has been dormant between the years 1910 and 1974, when it was revived by the French School in a series of papers by Cea, Murat, Zolesio and Pironeau. Most of the details on treatment of elliptic sustems can be found in the O. Pironeau monograph [16]. The original idea of Hadamard can be summarized as follows:

Around 1910, Hadamard proposed a scheme involving the perturbation of Green's function. Let $G(x,y,\Gamma_0)$ be the Green's function for the Laplace operator for a region Ω whose boundary is Γ_0.

Let Γ_ε be a perturbed boundary:

$$\tilde{x} \in \Gamma_\epsilon \quad \text{if } \tilde{x} = x + \epsilon\mathbf{n} \cdot \boldsymbol{\eta}(x), \quad \text{where}$$
$x \in \Gamma_0$, $\mathbf{n}$ – unit normal to Γ_0.
$\boldsymbol{\eta}(x)$-an arbitrarily chosen perturbing vector function.

Then,

$$\left.\frac{\partial G(x, y, \Gamma_\epsilon)}{\partial \epsilon}\right|_{\epsilon=0}$$

is the crucial 'generator' of such infinitesimal perturbation [17]. This idea of Hadamard was dormant until the early 1970s when it was revived by the French School.
If we follow Pironeau then the Γ-differentiability can be introduced. Let a functional $F(\Omega)$ depend on the shape of the region $\Omega \subset C$ (where C is a bounded region). We parametrize the boundary Γ of Ω and introduce

$$\Gamma_\alpha = \{x + \epsilon\alpha \cdot \mathbf{n}(x) : x \in \Gamma\}.$$

F is Γ differentiable if, for any

$$\eta(x) \in L_2(\Gamma),\ F(\Omega_{\epsilon\alpha}) - F(\Omega) = \int_\Gamma F' \cdot \eta \, ds + (\|\eta\|)$$

for some function F' that is dependent on η. (Γ_α is the boundary of the perturbed region $\Omega_{\varepsilon\alpha}$.) The optimality condition is analogous to the classical condition of the vanishing of the first variation of a functional. Pironeau goes in some detail over the problems involving the biharmonic equation, the Stokes problem, the incompressible Navier-Stokes equation:

$$-\nu\Delta u + u \cdot \text{grad}\, u = -\text{grad}\, p \text{ in } \Omega,\ \text{div}\, u = 0,$$

and devotes a brief discussion (of one page) to the eigenvalue problems. (See [16]).

Here the differentiability of the boundary restricts the applicability of this method to cases where smooth, or piecewise smooth shapes prevail. Unfortunately that is not the case with designs of plates or shells for optimum strength. Here we could introduce some techniques that are entirely new to this area, and in fact the author is the only researcher who apparently tried to pursue this approach.(see [18]). Whether some successive numerical studies will shortly emerge following this model-theoretic approach remains to be seen. But the same remark could be made about any new theory. We shall reserve this rather exotic approach for the last chapter. It requires mathematical sophistication and a knowledge of model theory that cannot be expected from even a sophisticated engineer or applied mathematician. For the time-being let us concentrate on Hadamard's approach and related matters. It is appropriate to mention several theoretical efforts by analysts who attempted to approach the perturbation of the domain problems by complex variable arguments, predominantly using conformal mapping techniques. For examples of this approach see Bergman and Schiffer [19], or Garabedian [20]. There are several basic difficulties. These techniques are developed only for two-dimensional problems. Also, the problem of "control" by a conformal map is not well posed. There exist well-known examples showing that, for example, very small perturbations of the boundary can produce large changes in the values of the solutions to the Dirichlet problem. Analytic functions exhibit bad behavior in this respect, thus making this approach to boundary perturbations difficult, and necessitating the checks on the dependence on the data. This is not a criticism of a technique that has been extremely successful in fluid flow problems, in classical theory of elasticity, and in the electramagnetic field theory. Examples of applications to fluid flow problems may be found in almost any advanced textbook on that subject. The remarks above are only intended as a warning and not as a complete negation of an established and very productive technique.

(5.10) Problems in shape optimization.

(5.10.1) Introduction

With advances in high speed computation a new technology was introduced to design of complex engineering systems. Most problems concern changes in design which can be regarded as changes in shape. However, the mathematical modelling introduces a fairly artifical problems whereby changes in some dimensions or directions are regarded as changes in values of functions while changes in other directions are regarded as changes in the domain of such functions. Such problems are mathematically more difficult and present an higher degree of difficulty. They are the topic of this section. We outline some of the approaches used by the French school and some ideas of Banichuk and his colleagues, as well as our own developments.

(5.10.2) An example of domain optimization.

Let us consider an abstract problem of domain optimization. Let the functional

$J = \int_{\Omega} f(u, u_{x_i}, x_i)\, d\underset{\sim}{x}$ define the cost functional

for our minimization problem of assigning the minimum value to J be choosing an admissible "state function"

$\underset{\sim}{u}(\underset{\sim}{x})$, where $\underset{\sim}{u} = \{ u_1, u_2 \ldots u_m$, $\underset{\sim}{x} = x_1, x_2 \ldots x_n\}$

A well-known heuristic derivation using the "del" formalism can be found in reputable mathematical texts, for example in Courant and Hilbert [27] , relating the first variation of J to the corresponding variation $d(\partial\Omega)$ in the shape of the boundary $\partial\Omega$ of the compact region $\Omega \subset R^n$.

The variation of the boundary denoted by $\delta(\partial\Omega)$

is accomplished by varying the variables δx_j in the direction of the normal to $\partial\Omega$, i.e. by considering the product $<\delta x_m, n> = <\delta x_j, n_j>$ where n_j is the component of the unit normal vector $\underset{\sim}{n}$ in the direction of the coordinate x_j.

The corresponding variation of J is given by the heuristically derived formula

$$(5.2.1)\quad \delta J = \sum_{k=1}^{m} \Big[\int_{\Omega} \Big\{ \Big[\frac{\partial f}{\partial u_k} - \sum_{i=1}^{n} \frac{\partial}{\partial x_i}\Big(\frac{\partial f}{\partial u_{k,x_i}}\Big)\Big] \cdot \Big[\delta u_k - \sum_{i=1}^{n} u_{k,x_i} \cdot \delta x_i\Big]\Big\}\, d\underset{\sim}{x}\Big]$$

$$+ \sum_{k=1}^{m} \Big[\int_{\partial\Omega} \Big\{ \sum_{i=1}^{n} \Big(\frac{\partial f}{\partial u_{k,x_i}} \cdot n_i\Big) \cdot \Big(\delta u_k - \sum_{i=1}^{n} u_{i,x_k}\Big)\Big\}\, d\underset{\sim}{s}\Big]$$

$$+ \int_{\partial\Omega} \Big\{ \sum_{i=1}^{n} (n_j \cdot \delta x_j)\, f(\underset{\sim}{x})\Big\}\, d\underset{\sim}{s},$$

where ds is the infinitesimal measure assigned to the n-1 dimensional hypersurface $\partial\Omega$.

We could derive a similar expression rigorously postulating invariance of the minimum of J under changes of variables that transform coordinates x_i into new coordinates ζ_i that are "close to x_i", however the formula (5.2.1) can be also derived completely rigorously if one assumes existence of all partial derivatives regarded as weak derivatives and square integrability of the integrands. The δ-symbols can be then replaced by Gateaux variations

$$\delta x_i = [(x_i + \varepsilon\eta_i) - x_i], \text{ etc.}$$

Various possibilities of simplifying the expression (5.2.1) are discussed in Banichuk's monograph [28]. For example, if we assume that J is the Lagrangian functional and J assumes an extreme value if u represents correctly the state variable, then the Weierstrass form

$$[\frac{\partial f}{\partial u_k} - \sum_{i=1}^{n} \frac{\partial}{\partial x_i} (\frac{\partial f}{\partial u_{k, x_i}})] \text{ vanishes on the}$$

kinematic manifold (or perhaps on the subset of admissible states U in the engineering version), and only the two terms containing boundary integrals in formula (5.2.1) remain.

This is exactly the approach suggested by Banichuk in [28].

(5.10.3) <u>The functional analytic treatment of a</u>
          <u>change of domain.</u>
Specifically, we shall restrict our discussion to either R^3 or R^2 but all of our arguments can be rewritten in R^n without any changes.
Let Ω_0 be a bounded (open) domain in R^n.
We assign a vector field **V** on Ω assigning a vector $v(\underset{\sim}{x})$ to every point $\underset{\sim}{x} \epsilon \Omega$.

We assume that V is continuous in Ω and solenoidal. We define the infinitesimal transformation : $\Omega_0 \rightarrow \Omega_{dt}$, that maps a point

$\hat{x}_0 \epsilon \Omega_0$ into a point $\hat{x}_{dt} = \hat{x}_0 + v(\hat{x}_0) dt$.

Point $\hat{x}_0$ is to be regarded as a vector from the

origin to $\hat{x}_0$.

Various definitions of differentiation can be offered. For example, the Eulerian derivative of the cost functional $J(\Omega)$ is given by

$$\lim_{\Delta t \rightarrow 0} \{ (J(\Omega_{\Delta t}) - J(\Omega_0))/\Delta t \} = J_V (\Omega_0)$$

If this derivative exists, then its value depends only on the vector field **V**.

In fact, if the functional J depends only on the domain Ω, then upper and lower semiderivatives

$$\limsup_{\Delta t \to 0} [\{J(\Omega_{\Delta t}) - J(\Omega_o)\}/\Delta t] = \bar{J}_V$$

and

$$\liminf_{\Delta t \to 0} [\{J(\Omega_{\Delta t}) - J(\Omega_o)\}/\Delta t] = \underline{J}_V ,$$

(if they exist) depend only on the vector field **V**.

If the functional $J_{\mathbf{V}}(\Omega)$ is continuous and linear in **V** then Riesz' theorem implies that it can be written as an inner product. $J_{\mathbf{V}}(\Omega) = \langle G(\Omega), V\rangle$ in the inner product space assigned to **V**.

It was shown by Zolesio [23] that regularity (C^k, $k > 1$) of the domain Ω implies several properties of the vector field G, or vector distribution $G(\Omega)$. Zolesio refers to G as the gradient of J. G is a vector distribution on R^n. A restriction α to $\partial\Omega$ is written as g_n. $G = \alpha\ (g_n(\Omega)\cdot \mathbf{n})$.

The Hadamard formula expresses $\bar{J}_V$ in terms of g_n

$$\bar{J}_V = \int_{\partial\Omega} (g_n(\Omega)\cdot V(o)\ \cdot \mathbf{n})\ ds.$$

The integral over $\partial\Omega$ may not exist in the Lebesgue sense and may be regarded as a bilinear form. However in many important examples this form is a "genuine integral" over $\partial\Omega$.

For details of this approach see J.P. Zolesio's thesis of 1979 [23], or his articles [24], [47], [26]. The main idea is quite easy to grasp intuitively. We let the material flow and watch the changing shapes and the rates of change of the cost functional along the flow lines (i.e. stream lines of the velocity field),

Example of the "speed method".

Consider a plate occupying a region $\Omega \subset R^2$ having thickness $h(x) \geq h_o > o$. The boundary conditions are

$$W = 0 \ , \quad \frac{\partial w}{\partial n} = 0 \qquad \text{on } \partial\Omega .$$

We wish to minimize either the maximum value of the shear stress $|\tilde{\tau}_{xy}|$ or the von Misses stress invariant

$$[(\sigma_{xx} + \sigma_{yy})^2 + 3\ (\sigma_{xx} - \sigma_{yy})^2 + 12\ \tau_{xy}{}^2]^{\frac{1}{2}} \ ,$$

or perhaps the first natural frequency.

Let $J_1(w,h) = \iint_\Omega \left(\frac{Eh}{2(1+\nu)}\ W_{xy}\right) dx\ dy$

We formally differentiate

$$J_1'(w,h) = \iint_\Omega \left(\frac{Eh}{2(1+\nu)}\ W_{xy}'\right)\cdot\theta_{m_i}\ dx\cdot dy$$

$$- \int_\Gamma \left\{\left(\frac{Eh}{2(1+\nu}\ W_{xy}'\right)\cdot\theta_{m_i}\right\} ds. \quad \left(\theta_{m_i} = \begin{cases} 0 \text{ if } x \notin m_i \\ 1 \text{ if } x \in m_i \end{cases}\right)$$

If m_i is an interior region, then the step function θ is identically zero on Γ.

Hence $J_1' = \iint \frac{E}{2(1+\nu} (\dot{w}_{xy})\ dxdy - \langle \text{grad } w, V\rangle\ d\Omega.$

Now, we need to experiment with velocity fields that produce a negative value of J'.

The meaning of $\dot{w}_{xy}$ should be clarified, since we can adopt either the Lagrangian or Eulerian point of view regarding changes in the values of functions along stream lines of the velocity field **V**.

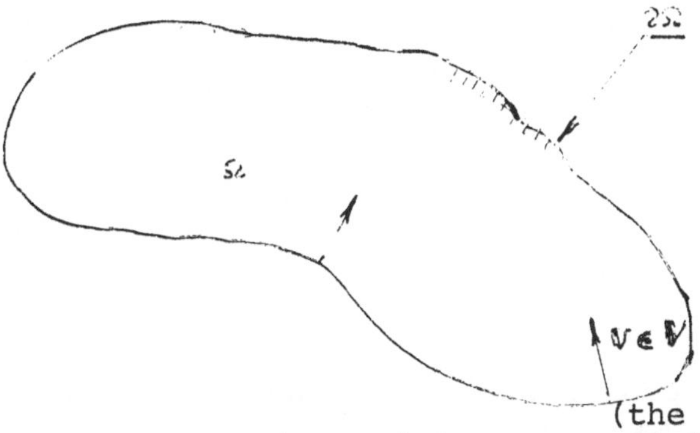

(the velocity field)

Figure 5.1

(5.10.4) Comment on "local" functionals

The optimization problems involving local criteria such as maximum deflection at a point or maximum stress are more difficult than problems with global optimality criteria such as compliance or total energy. While several techniques have been developed for handling numerically global optimization problems, basically there is only one way to initiate the theoretical investigation for perturbation of functionals involving integration over the domain (at least to the best of my knowledge). The recent techniques of the Russian school(Banichuk, Chernouško, Litvinov), French school(Cea, Zolesio, Roussellet) and University of Iowa program (Haug, Choi, Arora) developed several techniques for perturbing functionals depending on integration over the spatial domain, which involve fictitious variation of the domain. Such techniques generally fail in perturbation of local functionals. For example, maximum stresses most frequently occur on the boundary of the region defining the shape of an elastic object. If one uses "the speed" technique it is necessary to consider the displacement of the point of maximum stress. The usual formual is $\sigma' = \frac{d\sigma}{d\hat{x}} + < \text{grad}\ \sigma, V(\hat{x}_\tau) >_\Omega$

where $\hat{x}$ denotes the moving point of the region Ω, with $\hat{x}_\tau = \hat{x} + \tau\,\mathbf{V}(\hat{x})$, and with $\mathbf{V}$ denoting a fictitious velocity of points in the region Ω. This is generally useless in the investigation of local perturbations. K.K. Choi introduced an interesting variant where all terms in the sensitivity equation depend only on the boundary displacements thus bypassing the usual difficulties. Details are impossible to discuss without copying of rather lengthy formulas. The readers are referred to [22].

5.11 Introduction to simultaneous control optimization.

The discussion given below can be found in the article [46] of the author and V. Dannon. While the actual loads in the 'real world' always contain some random inputs, it is the highly unlikely dynamic load that may cause a disaster. The design of a structure or of a mechanical system to withstand such conspiracy of unlikely events has not been systemized and no consistent theory has been developed to the best of our knowledge. Some ideas have been advanced by Komkov [31] and Komkov and Coleman [32] but so far no consistent theory has emerged.

The basic problems outlined here are the same for a single member as for a large structure, and the difficulties can be illustrated by considering the optimal design of a single beam against random dynamic loads.

In order to formulate some algorithms to deal with this rather difficult problem, we establish a number of preliminary results.

Sensitivity is one of the most important considerations in any arguments concerning optimal or near optimal designs or optimal controls.

For example, the problem of finding an optimal control or optimal design has a necessary condition that the sensitivity of the system is zero at such optimal point.

Thus, if one establishes some properties of

the sensitivity function, it suffices to look for zeros of this function to determine all configurations, or all controls which are candidates for the adjective 'optimal'. However, in practical numerical applications, it suffices to look for low sensitivity to establish some near-optimal controls or near-optimal designs.

For this reason sensitivity arguments seem to dominate the numerical approaches in the present day optimization literature.

Sensitivity of optimal control is a popular topic of research. See, for example, the monograph of Dontchev [33], or [34].

Sensitivity to design changes has been the subject of an intensive recent investigations. The names of E. Haug, K. Choi, J. Taylor, N. Banichuk, J. Cea, T. Zolezzi, N. Olhoff, and P. Pedersen, are prominent in the recent theoretical developments. There is now a well-established body of results and corresponding numerical techniques for optimization of engineering systems. Some difficulties and inconsistencies of commonly used algorithmic procedures have been analysed and corrected. Our understanding of the whole area of computer-assisted design has been greatly enhanced in the last few years.

Sensitivity and optimal control of systems whose design is regarded as constant is perhaps one of the fastest growing areas of mathematics. The results attained in the last decade are too numerous and too diverse to catalogue in a research article.

Unfortunately, almost no literature apart from [46] exists on combining the two theories. Certainly, no articles came to our attention on the subject of simultaneous optimal design and optimal(extremal) excitation, that is, of dealing with the worst possible loads in attempting an optimal design of a complex mechanism or structure.

The synthesis of an optimal control algorithm for the behavior of worst dynamic loads against a specific design was discussed in [4]. Several serious difficulties exist. One difficulty discussed in Section 2 is the lack of convexity of the set of 'optimal' (worst possible) controls.

If $q_1(t,x)$, $q_2(t,x)$ are both 'optimal'(in this sense) controls, then $\lambda q_1 + (1 - \lambda)q_2$ is not optimal for any λ such that $0 < \lambda < 1(\lambda \neq 0, \lambda \neq 1)$. While optimal controls in structural dynamics are in general nonunique, this nonuniqueness is here combined with the lack of convexity in problems concerning the 'worst' excitation'. In turn, this causes a fundamental difficulty confronting the designer of an engineering system. If we wish to design a system which resists 'best' the 'worst' possible admissible loads, against which of the worst loads should we optimize? To demonstrate an inherent difficulty facing the class of problems, we prove a nonuniqueness theorem. The implications of this theorem are fairly subtle. It is not impossible to produce an optimal design, but the word 'optimal' has to be redefined.

Let us first discuss the background to our main result. It is in no sense complete or encyclopedic presentation. For such presentation of the state of the art in optimal design of structures around 1981 read the expository article of E.J. Haug [41]. No comparative exhaustive general review of the current research has been published since then.

We discuss only the techniques that are pertinent to our main result, that is, to the algorithm for simultaneous optimal design and 'worst adverse load control conditions'. Specifically, we consider separately the aspects of 'worst control' and of 'best design'. We include a suggestion of simultaneous treatment by considering the Pontryagin's characterization of the optimal control as a constraint applied to the best design procedure.

The proof of uniqueness(or rather nonuniqueness) and of existence based on convergence of approximate designs are left for Part 5. A pertinent review of some results concerning control and optimization of structures is necessary in any attempt at formulating a consistent theory of simultaneous control and design optimization.

Our first step is an introductory discussion

review of the functional analytic setting and of some basic mathematical concepts related to structural optimization and to the theory of optimal control.

5.12 A General Discussion of the Principles of an Optimal Design for an Optimally Controlled Distributed Parameter Structural System

(5.1.0) Notation and Functional Analytic Setting for a Bernoulli beam.

In this section we use common engineering notation. As before, we denote by

$w(x, t)$ displacement,
$A(x)$ cross-sectional area,
ρ material density (a positive constant),
$\mathscr{E}$ total energy ($\mathscr{E} = \mathscr{E}(t)$ is a nonnegative function),
u is the design (control vector),
L a linear differential operator,
$H_0^{2,1}$ the Sobolëv space which is the closure of C^∞ functions with compact support with respect to the norm

$$\|w\|_{H_0^{2,1}} = \left[\int_0^T \int_{-l}^{+l} [(w_{xx})^2 + (w_t)^2]\, dx\, dt\right]^{1/2}.$$

The corresponding energy norm is

$$|||w||| = \left[\int_0^T \int_{-l}^{+l} [\rho A(x) w_t^2 + EI(x) w_{xx}^2]\, dx\, dt\right]^{1/2}$$

The subscripts denote partial differentiation. Thus, w_t stands for $\partial w/\partial t$, w_{xx} for $\partial^2 w/\partial x^2$, etc.

$\langle u, v\rangle$ an inner product of two functions in some Hilbert space. If the space is not specified, then $\langle,\rangle$ denotes the usual L_2 products;

σ_{ij}, or τ_{ij} denotes a stress component, $q(x, t)$ a distributed load.

The design vector $\mathbf{u}$ is an element of a Banach space L_p, $p \geq 2$.

$\delta(x)$ the Dirac delta 'function'. $Y(x) \cdot \delta(x - x_0)$ has the obvious meaning if δ is the δ-distribution of Schwartz over the test space of C_0^∞ functions, whose support contains $[-l, +l]$, for any (smooth) function $Y(x) \in C_0^\infty$.

The functions $A(x)$ and $I(X)$ are square integrable and positive on the interval $[-l, +l]$.

h a variable (generally some dimension) that is a design parameter,

Q the space of admissible loads which is a subset of $H^{-2,-1}(\Omega)$. The norm in Q is the operator norm:

$$\|q\| = \max\{\langle q, w\rangle; \|w\|_{H_0^{2,1}} = 1\}.$$

The discussion given below is a repetition of the arguments of [46].

5.1.1 Sensitivity of Design Problems For Dynamically Loaded Beams

5.1.1.1 Introduction and Formulation of the Problems

We shall consider the following case. A random dynamic load is applied to a structure, or to a single member. The class of admissible loads Q contains either $L_2[-l, +l]$ distributed loads or a finite number of point loads of the form $\sum_{i=1}^{n} c_i\delta(x - x_i)$. The admissible loads $q(x, t) \in Q$ (the set of admissible loads) are such that $\|q(x, t)\| \leq 1$. In designing a structure, we shall assume that for each design $\mathbf{u}(x)$, the applied load $q(x, t) \in Q$ is either 'the worst possible' admissible load, or a load which is in some sense close to the 'worst possible load'. The load $q(x, t)$ may be exerted by mother nature in the form of a wind load, or an earthquake load, or it may be applied by casual traffic or other forms of nondeterministic input. We may regard the applied admissible load as a control which attempts to maximize the cost functional J on some preassigned time interval $[0, T]$, subject to certain constraints assigned to the state function. For a single beam, the state function $w(x, t)$ satisfies boundary conditions which are of the form

$$\text{(B1)}\quad w(\pm l, t) = 0, \qquad EI(u(x), x)\frac{\partial^2 w}{\partial x^2}\quad (\pm l, t) = 0$$

$$\text{(B2)}\quad w(\pm l, t) = 0, \qquad \frac{\partial w}{\partial x}(\pm l. t) = 0$$

$$\text{(B3)}\quad \left.\begin{aligned} &EI(u(x); x)\cdot\frac{\partial^2 w}{\partial x^2} = 0 \\ &\frac{\partial}{\partial x}\left(EI(u(x); x)\frac{\partial^2 w}{\partial x^2}\right) = 0 \end{aligned}\right\} x = \pm l,\ T \geq t \geq 0,$$

The initial conditions are either

(I1) $w(x,0) \equiv 0, \qquad w_t(x,0) \equiv 0, \quad x \in [-l, +l]$

or

(I2) $w(x,0) = w_0(x), \qquad w_t(x,0) \equiv 0, \quad x \in [-l, +l].$

The cost functional which the nature may extremize could be the maximum of the total energy $\mathscr{E}(t)$, $t \in [0, T]$, which is given by

$$\mathscr{E}(t) = \tfrac{1}{2}\int_{-l}^{+l}\left\{EI(u(x);x)\left(\frac{\partial^2 w(x,t)}{\partial x^2}\right)^2 + \right. \qquad 5.1.1a$$

$$\left. + \rho A(u(x);x)\left(\frac{\partial w(x,t)}{\partial t}\right)^2\right\} \mathrm{d}x.$$

That is, the nature applies an admissible control $\hat{q}(x,t)$ such that for some $\hat{t} \in [0, T]$

$$\mathscr{E}_{\hat{q}}(\hat{t}) \geqslant \mathscr{E}_q(t)$$

for all $q(t) \in Q$ and all $t \in [0, T]$. Here $\mathscr{E}_q(t)$ denotes the value of the total energy of the system at the time t, which is attained by applying an admissible control $q(x,t)$. The state of the system is given by the basic equation of motion

$$L(w;u) = \frac{\partial^2}{\partial x^2}\left(EI(\mathbf{u}(x);x)\frac{\partial^2 w(x,t)}{\partial x^2}\right) + \rho A(\mathbf{u}(x);x)\cdot\frac{\partial^2 w(x,t)}{\partial t^2}$$

$$= q(x,t), \qquad 5.1.2$$

with the derivatives considered as weak derivatives. The appropriate functional space assigned to w is the Sobolëv space

$$H_0^{2,1}([-l,+l]\times[0,T]).$$

We also consider the homogeneous equation

$$L(w,\mathbf{u}) = 0. \qquad 5.1.2H$$

Alternately, we may consider as our cost functional the maximum deflection of the beam:

$$D_q = \max|w(x,t)| \quad \begin{cases} t \in [0,T], \\ x \in [-l,+l], \end{cases} \qquad 5.1.1b$$

or the maximum stress

$$\bar{\sigma}_q = \max\left\{\left|\frac{M(x,t)\cdot c(x)}{I(x)}\right|\right\} \quad \begin{cases} t \in [0,T] \\ x \in [-l,+l] \end{cases}$$

Again, we assume that nature exerts an admissible control $\hat{q}(x, t)$ which maximizes D_q or $\bar{\sigma}_q$, respectively. We may wish to determine the effects of design on the value $D_{\hat{q}}$, $\bar{\sigma}_{\hat{q}}$, or $\mathscr{E}_{\hat{q}}$, respectively, so that the worst possible dynamic load has least effect on the maximum deflection, maximum stress or maximum value of the total energy of the beam, respectively.

The set of admissible designs is a suitable subset of $L_2[-l, +l]$. For example, in realistic designs we admit only piecewise continuous functions $\mathbf{u}(x)$ as admissible designs for obvious manufacturing reasons. However, for the sake of convenience we could admit any $L_2[-l, +l]$ function as an admissible design. In optimal design theory, 'chattering' functions may have to be considered, but this is not the case in our theoretical development concerning an improvement algorithm. Here, we are given a design $\mathbf{u}_0(x)$ and we wish to change it so as to improve the value of a cost functional. Hence, we can approximate 'chattering' designs with a suitably designated small error, by piecewise smooth functions.

5.1.2 The 'Worst' Control

The theory of 'worst' controls has not been developed successfully to the best of our knowledge even for criteria involving only the compliance or the total energy functionals. One cannot transplant results of optimal control theory with only suitable sign changes to derive analogous theorems for the 'worst' possible control, or the 'best' (optimal) excitation of a system.

The reason for this lack of symmetry is the nonconvexity of optimal excitations, compared to convexity of the optimal controls if one assumes a Banach space structure for admissible functions representing the forcing terms (controls).

This lack of convexity prevents us from asserting uniqueness theorems, such as the theorem on uniqueness of optimal finite state (see [36]). However, some theoretical results of optimal control theory can be applied to our case. We shall state our results in the form of lemmas and theorems. We assume that the initial design $\mathbf{u}_0(x)$ is given for each improvement algorithm.

We assume that the boundary conditions and initial conditions for the displacement function $w(x, t)$ are given by either (B1) or, (B2) or, (B3) and by either (I1) or (I2).

1.2.1. *The Design Problem for a Simple Vibrating Column*

We identify the 'best' or 'worst' loads by their effect on stability of the column, or on the natural frequency of vibration. Thus, we formulate the extremal problem. With each admissible design $\mathbf{u}(x) \in U$, find the worst possible admissible load $\hat{q}(x, t) \in Q$, such that the first natural frequency of the vibrating column assumes the lowest possible value. We seek an admissible design $\hat{\mathbf{u}}(x) \in U$ which is 'best' in this sense.

$$\lambda_1(\hat{\mathbf{u}}) = \max_{\mathbf{u}\in U}[\min_{\mathbf{q}\in Q} \lambda(\mathbf{u}, \mathbf{q})].$$

We could offer a similar definition for a static deflection case; however, if we identify λ with the lowest buckling load, we reduce the problem to that of simple optimization of the design, that is to a standard version of finding the optimal shape of a column. We shall consider only the more difficult dynamic case.

Some Preliminary Results.

The following lemmas are given for the sake of completeness.

LEMMA 1. *We assume the initial conditions* (I1). *Let J denote a cost functional which is either* (1.1a), (1.1b) *or* (1.1c). *Let* $[0, \tau]$ *be a subinterval contained in* $[0, T]$. *Let* $\mathbf{q}_\tau(x, t)$ *be an admissible load (implying* $\mathbf{q}_\tau(x, t) \in H^{-2,-1}$ *and* $|\mathbf{q}_\tau| \leq 1$) *which maximizes* $J_{\mathbf{q}}(t)$ *at* $t = \tau < T$. *Then there exists a control* $\hat{\mathbf{q}}(x, t) \in Q$ *such that* $J_{\hat{\mathbf{q}}}(T) \geq J_{\mathbf{q}_\tau}(\tau)$.

Proof. Choose $\hat{\mathbf{q}}(x, t) \equiv 0$ on the interval $[T - \tau, T]$ and

$$\hat{\mathbf{q}}(x, t) = \mathbf{q}_\tau(x, t) \quad \text{on } [0, \tau].$$

Then

$$J_{\hat{\mathbf{q}}}(T) = J_{\mathbf{q}_\tau}(\tau).$$

Lemma 1 indicates that in all problems involving initial conditions (I1) it suffices to extremize the cost functional J at the final time $t = T$.

LEMMA 2 (Duhamel's principle). *Let* $w(x, t)$ *denote the solution of Equation* (1.2). *Then there exists an* $L_2(\bar{\Omega} \times \bar{\Omega})$ *function* $G(x, \xi, t, \tau)$ *which depends only on the boundary conditions, but is independent of* $q(x, t)$, *or of the initial conditions such that for any* $t = \bar{t}$, *the following 'variation of parameters' formula is applicable*

$$w(x, \bar{t}) = \int_0^{\bar{t}} \int_0^l G(x, \xi, t, \tau) \cdot q(\xi, \tau)\, d\xi\, d\tau + w_H(x, \bar{t}), \qquad 5.2.1$$

where $w_H(x, t)$ *denotes the (unique) solution of the homogeneous equation* (1.2H) (*i.e. with* $q(x, t) \equiv 0$) *while* $\bar{\Omega}$ *denotes* $[-l, +l] \times [0, T]$.

Proof. The proof of this version of Duhamel's principle exists in the literature (also see [36] for explanations regarding its use), but a word of caution should be inserted. It is valid for the class of admissible loads specified above and for the boundary and initial conditions stated above. It is generally false if certain improper boundary conditions are assigned.

LEMMA 3. *We assume that the initial displacement* $w_0(x)$ *of the beam (in the initial condition* (I2)) *is such that an admissible static load* $q_s(x)$ *can be found*

($q_s(x, t_1) = q_s(x, t_2)$ for all t_1, $t_2 \in [0, T]$) such that the static deflection $w_0(x)$ is maintained. By this statement we mean the existence of a solution of (1.2), $\bar{w}(x, t)$, *such that*

$$\frac{\partial}{\partial t}\bar{w}(x, t) \equiv 0 \text{ on } [0, T] \quad \text{and} \quad \frac{\partial^2}{\partial x^2}\left(EI(x)\frac{\partial^2 \bar{w}}{\partial x^2}\right) \equiv q_s(x).$$

Then the conclusion of Lemma 1 *is true if the initial conditions* (I1) *are replaced by the conditions* (I2).

Proof. Let q_τ be the extreme control on $[0, t] \subset [0, T]$. We construct an admissible control $\hat{q}(x, t)$, $t \in [0, T]$ as follows

$$\hat{q}(x, t) = q_\tau(x, t), \quad t \in [0, \tau]$$

$$\hat{q}(x, t) = q_s(x) \quad \text{on } [\tau, T].$$

Using Duhamel's principle (2.1) we conclude (after a rather easy manipulation) that $J_{q_\tau}(T) = J_{\hat{q}}(T)$.

Comment. In Lemmas 1 and 3 we only needed to show that we can do at least as well on $[0, T]$ as on $[0, \tau] \subseteq [0, T]$ in worsening the value of the cost functional. It is not hard to show that whatever damage could be done on $[0, \tau]$ by admissible controls, worse damage can be done on a longer interval of time. We conclude that it suffices to examine the 'worst possible' conditions at the end point T of any fixed time interval $[0, T]$.

5.1.2.3

THEOREM 1 (Pontryagin's principle).

Note. This version of Pontryagin's maximality principle was proved by the author [36] It is included here for the sake of completeness.

Let $\hat{q}(x, t)$ denote the extremal excitation, $\hat{w}(x, t)$ the corresponding displacement of a beam satisfying either (B1), (B2) *or* (B3) *at each end $x = \pm l$ and either* (I1) *or* (I2) *at $t = 0$. $w_H(x, t)$ denotes a solution of the homogeneous equation* (1.2H) *satisfying the set of 'worst possible' final conditions*

$$w_H(x, T) = \bar{w}(x, T), \qquad \dot{w}_H(x, T) = \dot{\bar{w}}(x, T),$$

(w_H, $\dot{w}_H$ *assign the largest possible value to the cost functional J at the time $t = T$). Then the energy of the system is maximized at the time $t = T$ if*

$$\int_{-l}^{+l} \hat{q}(x, t)\dot{w}_H(x, t)\,dx \geq \int_{-l}^{+l} q(x, t)\dot{w}_H(x, t)\,dx, \qquad (5.2.2)$$

for all $t \in [0, T]$ and for any admissible control $q(x, t) \in Q$.

Remarks. Pontryagin's principle for the functional D_q is more difficult to derive. The convenient L_2 arguments cannot be applied to the functional representing the L_∞ norm of $w(x, T)$. We observe that the difficulty arises from our inability to locate the point of maximal deflection at the time $t = T$. For a statically determinate beam the deflection at a point $x_0 \in [-l, +l]$ is given by the Betti–Castigliano formula

$$w(x_0) = \int_{-l}^{+l} \left(\frac{M(x, t) \cdot m(x, x_0, t)}{EI(x)} \right) \mathrm{d}x + w_H(x, t)|_{x=x_0}, \qquad 5.2.3$$

where $M(x, t)$ is the moment at the point x at time t; $m(x, x_0, t)$ is the moment at x caused by application of the Dirac delta function at x_0, and $w_H(x, t)$ is the solution of the homogeneous equation satisfying the appropriate initial and boundary conditions for the problem.

The advantage of this formula for a statically determinate case stems from the fact that $M(x, t)$ and $m(x, x_0, t)$ are design independent. This feature resulted in many results in sensitivity analysis and optimization of a deflection at a given point for purely statically determinate cases, or cases which can be reduced to a purely static analysis.

In maximizing the strain energy at the time $t = T$ the following form of Pontryagin's principle can be derived by techniques resembling those given in [36] or [46].

THEOREM 1a. *For all $t \in [0, T]$ the extreme control $\hat{q}(x, t)$ satisfies the relation*

$$\int_{-l}^{+l} ([\hat{q}(x, t) - \rho A\ddot{\hat{w}}(x, t)]\dot{w}_H(x, t))\, \mathrm{d}x \geq \int_{-l}^{+l} ([q(x, t) - \rho A\ddot{w}(q; x, t)]\dot{w}_H(x, t))\, \mathrm{d}x$$

for any admissible control $q(x, t)$ and corresponding admissible displacement $w(q; x, t)$. $w_H(x, t)$ has the same meaning as before.

No corresponding form of maximality principle has been developed for the functionals D_q and $\bar{\sigma}_q$. This makes corresponding development of sensitivity analysis very difficult. In particular, the technique developed by W. Prager [37] that is making use of the independence of design from the bending moment and the shear load, has so far escaped generalizations to statically indeterminate cases in which Prager's assumptions are false. However, the energy arguments permit us to perform sensitivity analysis for dynamic cases if the cost functional is either the compliance functional, the total energy, or the strain energy functional.

5.1.3 Sensitivity Analysis

We consider the basic equation of state

$$L(u)w - \hat{q} = \rho A(\mathbf{u}) w_{tt} + (EI(\mathbf{u}) w_{xx})_{xx} - \hat{q}(\mathbf{u}, x, t) = 0, \quad (5.3.1)$$

with boundary conditions (B1) or (B2) and initial condition (I1). An equality constraint $\int_\Omega \phi(\mathbf{u})\,\mathrm{d}x = 0$ is imposed on the design variable $u(x, t)$. (For example, we wish to keep constant weight.) The forcing term $\hat{q}(u; x, t)$ satisfies 'the worst possible' requirement expressed in the language of Pontryagin's principle

$$\int_{-l}^{+l} \int_0^T \hat{q}(\hat{u}; x, t) \cdot \frac{\partial w_H(\hat{\mathbf{u}}, x, t)}{\partial t}\,\mathrm{d}t\,\mathrm{d}x - \hat{C}(w(\hat{\mathbf{u}}), T)$$
$$= \int_\Omega \phi(w)\,\mathrm{d}x = \psi(w) = 0 \quad (5.3.2)$$

where Ω is the open interval $(-l, l)$ and where $\hat{C}(w(\hat{\mathbf{u}}), T)$ is the largest possible (finite) energy level attainable by means of admissible controls. The maximal condition $\mathscr{E}(w(x, T) - C(w, T) = 0$ can be replaced by Pontryagin's necessary condition namely $\psi(w) = \int_\Omega \phi(w)\,\mathrm{d}x = 0$. Thus, Pontryagin's principle is used as a constraint.

Introducing the adjoint variable $\lambda(x, t)$ we write Equation (3.1) in the variational form

$$\int_{-l}^{+l} \lambda(x, t) \cdot (L(\mathbf{u})w - q)\,\mathrm{d}x = 0. \quad 5.3.3$$

We require that $\lambda(x, t)$ satisfies the final condition $\lambda(x, T) \equiv 0$, and that

$$L^* \lambda(x, t) = \frac{\partial \psi}{\partial w}, \quad 5.3.4$$

where L^* is the adjoint of L.

We acknowledge the fact that the state equation (3.1) remains valid if we change the design. We use formally the engineering 'δ' notation:

$$\delta\{\langle \lambda(x, t), (L(\mathbf{u})w - \hat{q}(\mathbf{u}))\rangle_\Omega\} = 0, \quad 5.3.5$$

where

$$\langle f, g\rangle_\Omega \equiv \int_{-l}^{l} (f \cdot g)\,\mathrm{d}x.$$

This may be written in the form

$$
\begin{aligned}
0 &= \delta\langle\lambda, L(\mathbf{u})w\rangle - \delta\langle\lambda, \hat{q}(\mathbf{u})\rangle \\
&= \langle\lambda, \{\delta L(\mathbf{u})\}w\rangle + \langle\lambda, L(\mathbf{u})\,\delta w\rangle - \langle\lambda, \delta\hat{q}(u)\rangle \\
&= \langle\lambda, R_u\rangle\,\delta u + \langle L^*\lambda, \delta w\rangle - \left\langle\lambda, \frac{\partial\hat{q}}{\partial\mathbf{u}}\right\rangle \delta\mathbf{u},
\end{aligned}
$$

where

$$
R_u = \rho\frac{\mathrm{d}A}{\mathrm{d}\mathbf{u}}\,w_{tt} + \left(E\frac{\mathrm{d}I}{\mathrm{d}\mathbf{u}}\,w_{xx}\right)_{xx} - \frac{\partial\hat{q}}{\partial\mathbf{u}}.
$$

Now,.

$$
\langle L^*\lambda, \delta w\rangle = \left\langle\frac{\partial\psi}{\partial w}, \delta w\right\rangle_\Omega = 0.
$$

Hence

$$
\langle\lambda, R_{\mathbf{u}}\rangle_\Omega - \left\langle\lambda, \frac{\partial\hat{q}}{\partial\mathbf{u}}\right\rangle_\Omega = 0.
$$

The joint sensitivity corresponding to the functional J subject to constraints of our problem is given by

$$
\Lambda = \langle\lambda, R_{\mathbf{u}}\rangle_\Omega + \mu_1\frac{\partial J}{\partial\mathbf{u}} + \mu_2\frac{\partial}{\partial\mathbf{u}}\int_0^T (\hat{q}\dot{W}_H\,\mathrm{d}t - \hat{C}). \quad (5.3.6)
$$

μ_1, μ_2 are Lagrangian multipliers. To determine their values one has to set

$$
\int_{-l}^{+l}\phi(\mathbf{u})\,\mathrm{d}x = 0, \qquad \int_0^T (\hat{q}\dot{w}_H - \hat{C})\,\mathrm{d}t = 0, \quad \Lambda = 0,
$$

and solve for μ_1 and μ_2.

Of course, this is frequently easier said than done, and in large structural or mechanical systems the calculations may be quite lengthy. However, in principle, there are no fundamental theoretical difficulties associated with such computation, or with the design of corresponding computer software.

The main point of our analysis is not replacement of all iterative computations by a 'magic' single formula, but rather the construction of a numerical approach that combines the purely theoretical approach with some working iterative techniques in such manner that the most tedious numerical steps are avoided. More specifically, we want to avoid the worst features of the 'brute force' approach in which the derivatives, the state of the system, the worst possible load for that state, the sensitivity, etc. have to be recomputed at each iterative step. In particular, we wish to avoid repetitive computations of the state.

The term $\langle \lambda, R_u \rangle_\Omega$ can be determined by numerical techniques. The difficult term in our analysis is

$$\frac{\partial}{\partial \mathbf{u}}\left(\int_0^T (\dot{q}w_H - \hat{c})\, \mathrm{d}t\right),$$

where $\hat{c} = (1/T)\hat{C}$.

Specifically, $\hat{C}$ is design dependent. If we make a simplifying assumption that one set of worst possible finite conditions is established for all designs, i.e., that $\hat{C} = \mathscr{E}(T)$ is design independent, then dependence of $\dot{q}w$ on design can be easily established numerically. Let $\mathbf{w}_H$ denote the vector

$$\mathbf{w}_H = \begin{bmatrix} w_H \\ \rho A \dot{w}_H \end{bmatrix}$$

Then $\mathbf{w}_H$ solves the homogeneous equation

$$\dot{\mathbf{w}}_H = \mathscr{A}\mathbf{w}_H, \tag{5.3.7}$$

where

$$\mathscr{A} = \begin{bmatrix} 0 & (\rho A)^{-1} \\ -\dfrac{\partial^2}{\partial x^2}\left(EI(x)\dfrac{\partial^2}{\partial x^2}\right) & 0 \end{bmatrix}, \tag{5.3.8}$$

and satisfies the finite condition

$$\mathscr{E}(\mathbf{w}_H)\big|_{t=T} = \hat{C}. \tag{5.3.9}$$

That is $\mathbf{w}_H(t = T)$ and $\dot{\mathbf{w}}_H(t = T)$ are given, and w_H satisfies the backward evolution equation.

The sensitivity of the vector $w_H(x, t)$ to design changes is given by the operator equation

$$\frac{\partial \mathbf{w}_H}{\partial \mathbf{u}} = \left((-t + T)\frac{\partial \mathscr{A}}{\partial \mathbf{u}}\right)\mathbf{w}_H, \tag{5.3.10}$$

where $\partial\mathscr{A}/\partial u$ denotes the operator

$$\begin{bmatrix} 0 & \rho A^{-2}(u)\cdot A'(u) \\ -\dfrac{\partial^2}{\partial x^2} EI'(u)\dfrac{\partial^2}{\partial x^2} & 0 \end{bmatrix}, \tag{5.3.11}$$

where, as before,

$$A'(u) = \frac{\mathrm{d}A(u)}{\mathrm{d}\mathbf{u}}, \qquad I'(u) = \frac{\mathrm{d}I(u)}{\mathrm{d}\mathbf{u}}.$$

The form of $\hat{q}(x, t)$ is easily established numerically for a given $w_H(x, t)$.

$$\hat{q}(x, t) = \delta(x - \bar{x}) \cdot \dot{w}_H(\bar{x}, t), \tag{5.3.12}$$

where

$$|\dot{w}_H(\bar{x}, t)| = \max_{x \in \Omega} |\dot{w}_H(x, t)|.$$

It is not hard to show that the load $\hat{q}(x, t)$ is admissible for any $t \in [0, T]$ if $\delta(x - \bar{x}) \cdot w_H(\bar{x}, T)$ is an admissible (worst possible) load.

The sensitivity of $\langle \dot{w}_H, \hat{q}(x, t) \rangle$ is given approximately by

$$\frac{\partial \dot{w}_H}{\partial u} \left(\hat{q}(x, t) + \left[\int_{-l}^{+l} (\dot{w}_H)^{p-1} \, dx \right]^{1/p} \right), \tag{5.3.13}$$

where $\dot{w}_H \in L_\infty$ has been replaced by $\dot{w}_H \in L_p$ for some integer $p > 2$. This is a procedure which may lead to grave errors if w_H has more than one maximum and the values of the maxima of $|\dot{w}_H|$ are close. Substitution of L_p norm (even with p large) may cause us to miss a switching point (on the interval $[0,T]$) for the control $\hat{q}(x,t)$. We comment that, nevertheless, the device of replacing L_∞ norm by L_p norm, $p >> 1$, is in common use. See, for example, N.V. Banichuk [38], chapter 1, pages 30-35. Banichuk obtained some very reliable sensitivity and optimization results using this technique. (See [38,39].) However, the author is uneasy regarding its use unless some a priori conditions can be assumed concerning the cost functional.

5.1.4. Numerical Determination of the Worst Possible Finite State For a Given Design.

We use the technique described in [36] in which instantly optimal control was used to derive approximately optimal finite state. The technique for attaining approximately worst state is described below.

We partition the interval $[0, T]$ into a n-subintervals $[0, T_1]$, $[T_1, T_2] \ldots [T_{n-1}, T]$, with $T = T_n$.

We apply an arbitrary admissible control $\hat{q}_0$ at the time $t = 0$ (say $\delta(x)$) to the beam $x \in [-l, +l]$ and solve the equation $Lw_0 = \hat{q}_0$ finding the point $\bar{x}_1$ of maximal velocity $|\dot{w}_0(\bar{x}_1, T_1)| \geqslant |\dot{w}_0(x, T_1)$ for all $x \in [-l, +l]$. We then apply the control $\delta(x - \bar{x}_1)$ sign $\dot{w}_0(\bar{x}_1, T_1)$ to solve the equation $Lw_1 = \hat{q}_1$ on $[T_1, T_2]$ using $w_0(x, T_1)$, $\dot{w}_0(x, T_1)$ as the set of initial conditions, obtaining the point $\bar{x}_2$ of maximum velocity $|\dot{w}_1(\bar{x}_2, T_2)| \geqslant |\dot{w}_1(x, T_2)|$, for all $x \in [-l, +l]$. Then apply the control $\hat{q}_2 = \delta(x - \bar{x}_2)$ sign $\dot{w}_1(\bar{x}_2, T_2)$ to solve $Lw_2 = \hat{q}_2$ with $w_1(x, T_2)\dot{w}_1(x, T_2)$, given on $[T_2, T_3]$ etc...

LEMMA. *The control $\hat{q}$ maximizes the rate of energy increase*

$$\dot{\mathscr{E}}(w) = \frac{\mathrm{d}}{\mathrm{d}t}\int_{-l}^{+l} [EI(w_{xx})^2 + A\rho(w_t)^2]\,\mathrm{d}x \qquad (5.4.1)$$

on the interval $[0, T]$ *in the limit as* $n \to \infty$ *and* $(T_{i+1} - T_i) \to 0$. *That is* $\dot{\mathscr{E}}(\hat{w}(\hat{q})) \geqslant \dot{\mathscr{E}}(w(q))$ *for any admissible control* $q(x, t)$ *and corresponding displacement* $w(q(x, t))$.

The proof is fairly easy and is omitted.

5.1.5. An Example of A Possible Application To Electrical Networks

Let us digress by pointing out that the entire theory is applicable to other engineering systems and to other fields.

A random eletromagnetic pulse is applied to a network or to a segment of it. The class of admissible pulses Q is described similarly to the class of admissible loads and we assume that for each design $u(x)$, the applied pulse $q(x, t) \in Q$ is 'the worst possible admissible pulse'.

While this discussion seems to repeat the results of structural analysis, there are some essential differences. The equations of elasticity and structural analysis, such as the beam or plate equation, are generally of the 'mixed' type. The equations describing propagation of electromagnetic pulses are generally hyperbolic and results cannot be routinely 'translated'.

The pulse may be exerted in the form of an atmospheric shock or a lightning storm or it may be applied by casual transmission or as a by-product of an explosion etc.

We may regard the applied admissible pulse as a control which attempts to maximize the cost functional J on some preassigned time interval $[0, T]$ subject to certain constraints assigned to the state function. For a single dipole the state function is the four-dimensional vector potential.

$$\mathbf{W}(x, t) = \{w_1(x, t), w_2(x, t), w_3(x, t), w_3(x, t), w_4(x, t)\}.$$

It satisfies boundary conditions and initial conditions similar to the mechanical ones. $I(u(x), x)$ is then the impedance of the cross-section.

The cost functional J may be the rate of the total electromagnetic energy given by

$$\int_{-l}^{l}\left\{\frac{\partial}{\partial t}(\epsilon E^2+\mu H^2)+\operatorname{div}(\mathbf{E}\times\mathbf{H})\right\}\mathrm{d}x$$

where

$$\begin{aligned}\mathbf{E}(x,t)&=-\nabla w_1-\frac{\partial}{\partial t}(w_2,w_3,w_4)\\&=\left(-\frac{\partial w_1(x,t)}{\partial x}-\frac{\partial w_2(x,t)}{\partial t},\frac{\partial w_3(x,t)}{\partial t},-\frac{\partial w_4(x,t)}{\partial t}\right),\\ \mathbf{H}(x,t)&=\frac{1}{\mu}\nabla\times(w_2,w_3,w_4)\\&=\frac{1}{\mu}\left(0,-\frac{\partial w_4(x,t)}{\partial x},-\frac{\partial w_3(x,t)}{\partial x}\right),\end{aligned}$$

that is

$$\int_{-l}^{l}\frac{\partial}{\partial t}(\epsilon E^2+\mu H^2)\,\mathrm{d}x-[\mathbf{E}\times\mathbf{H}]_{-l}^{l}$$

represents the energy rate of change.

Applying variational principles, one can derive the telegraph equation describing the state of the system. Alternatively, J may be the maximal potential induced in the dipole for $t\in[0,T]$. The design problem for a dipole is to find an admissible design $\mathbf{u}(x)$ that will minimize the destabilizing effect of any admissible shock, that is, will limit the frequency of oscillations to the first resonance frequency of the dipole.

The analysis of the preceding sections applies to the design optimization against the worst possible electromagnetic shocks, as well as to the main subject of this paper, that is to optimization of mechanical and structural dynamic systems.

5.1.6. Concluding Remarks

So far, 'largeness' or 'complexity' has not been emphasized. In practice, it is of paramount importance.

Some fundamental difficulties, and corresponding numerical 'tricks of the trade' are closely related to the 'largeness of the systems', that is in Richard Bellman's terminology to the 'curse of large numbers'.

We should also comment on transition to large systems. A very large literature exists on the topic of matrix techniques and numerical approaches to the stress analysis of very large structures or networks (Zienkiewicz and Campbell [29], Haug and Arora [40], or chapter 1 of Haug et al. [44] contain an exposition of some of the commonly adopted techniques of analysis for large structural systems). Larger scale finite element approximations constitute a highly developed and specialized topic that deserves a systematic treatment from the point of view of 'worst control'.

It is tempting to outline some problems that are specific to large systems and to derive corresponding 'worst' control and design principles. However, any reasonable discussion of this problem would digress too far from the main theme and should be reserved for a separate research work.

Some comments should also be made about controllability.

The problem of controllability did not arise, for the simple reason that we have assumed an instant response of the structure. This is clearly false if the optimal control approaches a 'chattering' state, i.e. if the magnitude or direction of the control vector **U** has discontinuities which are finite in number in a finite interval of time, but the time interval separating two such discontinuities becomes too small for a feasible physical implementation. Theoretically, the structure cannot be controllable if any time interval between successive discontinuities of V is smaller than twice the time necessary for the compressive (tensile) stress wave to travel from the point of application of the control to the most remote point of the structure. Denoting the successive 'switching' time instances (i.e., times corresponding to successive discontinuities in control $u(x, t)$ by τ_1, τ_2, we must have $|\tau_2 - \tau_1| > 2\{\bar{l}/\sqrt{(E/\rho)}\}$ where $\bar{l}$ is the maximum length of the wave path through the structure, while ρ denotes the density of the material. $\sqrt{E/\rho}$ is the velocity of the compression wave in the ideal elastic material.

Computational Algorithms

Theoretical Foundations of An Optimal Design Algorithm. A Counter-Example

The most obvious 'practical' algorithm for obtaining a 'best' design against the 'worst' control would proceed as follows. Let us choose some initial design $u_0 \in U$. Using one of the established numerical techniques of control theory, let us design one of the 'worst' controls. For example, one can use the locally optimal control algorithm offered in the monograph [36] for a fixed design u_0, and arrive at some control $q_1(x, t)$ which is in some sense worst on a time interval $[0, T]$. Using a fixed control $q_1(x, t)$ we proceed to optimize the design and arrive at the design $u_1(x)$. Now, we proceed again to improve the control, arrive at a control $q_2(x, t)$, and so on.

Supposedly this process converges to a pair $(\tilde{q}(x, t), \tilde{u}(x))$. Unfortunately, as the following counter-example shows, that may not be the case.

Let us consider a single beam with very simple design criteria.

The beam is simply supported. As shown on Figure 1 the beam is manufactured by welding three sections of length $l/3$, each section has a constant rectangular cross-sectional shape. The width is constant. The height h is the design variable. In effect, we can choose heights h_1, h_2, h_3 for each of the beam segments. We impose a constraint of maximum weight W. That is,

$$(g\rho bl/3)\cdot(h_1+h_2+h_3)\leq W \quad \text{or} \quad h_1+h_2+h_3\leq K,$$

where $K = 3W/g\rho bl$. Only two sizes of height are available

$$h = H_1 = \frac{1}{2}K, \qquad h = H_2 = \frac{1}{4}K.$$

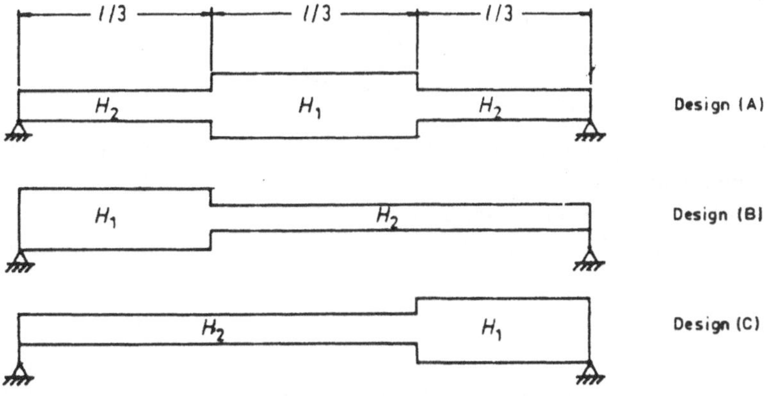

Fig. 1.

This limits the admissible designs to three configurations (A), (B), (C), as shown. The cost functional is the maximum displacement at the time T, where T is the period of vibration corresponding to the natural frequency in the configuration (B) or (C). Let the initial design be the design (B). The admissible controls $q(x, t)$ are any applied loads which consist of a distributed load. (i.e., bounded $L_2[0, l]$ functions of the spatial variable x), or Dirac delta function of x, and are piecewise continuous, bounded functions of t. Thus, $q \in Q$ implies that $q(x, t)$ is of the form $q(x, t) = f(x)\sigma(t)$, with $\|q\| \leqslant 1$, where

$$f(x) = f_1(x) + f_2(x), \qquad f_1(x) \in L_2[0, l], \qquad f_2(x) = \sum_{i=1}^{n} c_i\delta(x - \xi_i),$$

$$0 < \xi < l, \quad \sum |c_i| \leqslant 1, \qquad i = 1, 2, \ldots, n.$$

and $\sigma(t)$ is piecewise continuous, bounded function and $|q(x, t)| \leqslant 1$ for all $x \in [0, l]$, $t \in [0, T]$. We derive the 'worst' control $q_1(x, t)$ which is in some sense worst on a time interval $[0, T]$. Using a fixed control $q_1(x, t)$ we proceed to optimize the design and arrive at the design $u_1(x)$. Now, we proceed again to improve the control, arrive at a control $q_2(x, t)$, and so on.

Supposedly, this process converges to a pair $(\bar{q}(x, t), \bar{u}(x))$. The cost functional is the maximum displacement at the time T, where T is the period of vibration corresponding to the natural frequency in the configuration (B) or (C). Let the initial design be the design (B).

After an elementary computation, we derive the worst possible control to be

$$q_1(x, t) = \delta(x - \xi)\sigma(t),$$

where

$$\sigma(t) = \text{sign}(W(\xi_1, t)),$$

$$\xi_1 = \frac{l}{2} + l\left(\frac{1}{2} - \frac{1}{\sqrt{3}}\right), \qquad W(\xi_1, t) = C_1 \sin\left(\frac{2\pi t}{T}\right),$$

where C_1 is a positive constant which is easily computed, but its magnitude is of no importance.

Now we redesign the beam to improve the cost functional. The best design available is (C). The worst admissible load is now $q_2(x, t) = \delta(x - \xi_2)\sigma(t)$ with

$$\xi_2 = \frac{l}{2} - l\left(\frac{1}{2} - \frac{1}{\sqrt{3}}\right).$$

The optimal design against this control load is (B). Now $q_3(x, t) = q_1(x, t)$ and we have an infinite cycle, where no admissible design and no admissible control load satisfies the simultaneous extremality condition.

This counter-example was very simple. However, a similar phenomenon (of an emergence of an infinite cycle) has been observed in an attempt to simultaneously optimize the design and control in a more complex case. This counter-example indicates clearly that any sequential algorithm for simultaneous control and optimization should be viewed with suspicion. We point out that some of the worst features of an optimal excitation (i.e., of maximization of energy, deflection, stress or other cost functional) were not used in construction of the counter-example. The nonuniqueness, or the lack of convexity of optimal excitations was not even mentioned.

Since the most obvious numerical technique does not work, we need to design an alternate algorithm that avoids the infinite cycle phenomenon and produces at least a convergence to a near-optimal design under 'worst' admissible control.

5.2.2. Justification of the 'Simultaneous' Algorithm (S-Algorithm)

The 'simultaneous' method numerical approach consists of iteration introduced in the Section 1.3. Let the initial design be $\mathbf{u}_0$. The initial control $\mathbf{q}_0$ obeys the 'optimality' criterion:

$$q_0 \in Q \quad \text{and} \quad \int_\Omega (q_0 \cdot \dot{W}_H(q_0)\,\mathrm{d}x\,\mathrm{d}t = \max_{(\tilde{q}\in Q)} \left\{ \int_\Omega \tilde{q} \quad \dot{W}_H(\tilde{q})\,\mathrm{d}\mathbf{x}\,\mathrm{d}t \right\}$$

or at least approximates this 'optimality' criterion by some $\epsilon > 0$.

LEMMA 1. *Let Q be a bounded, convex subset of a Banach space. For a fixed design $\mathbf{u}_0 \in U$ the number*

$$\dot{C} = \sup_{(\tilde{q}\in Q)} \left\{ \int_\Omega \tilde{q} \cdot \dot{W}_H(\tilde{q})\,\mathrm{d}x\,\mathrm{d}t \right\}$$

exists.

Note. We deliberately omitted 'closed' in the hypothesis.

Proof. Without any loss of generality we assume Q to be a subset of the unit ball of $L_1(\Omega)$, thus including the δ-convergent sequences as elements of Q. We can use linearity and Duhamel's principle to conclude that

$$\max_{\substack{x\in[0,\, l]\\ t\in[0,\, T]}} \{|W(q(x, t); x, t)|\} = |\tilde{W}(\hat{x}, \hat{t})| \leq \tilde{M} \sup|q(x, t)| = \tilde{M},$$

for some number $\tilde{M}$, since $q \in Q$ is in a subset of the unit ball in $L_1(\Omega)$.

Since every bounded set of real numbers has a least upper bound, the existence of $\hat{C}$ is proved. While an optimal control in a space L_1, or L_∞ may not exist, it is not hard to show that for any $\epsilon > 0$ we can produce a smooth (i.e., C^∞) control which is within an ϵ of the number $\hat{C}$ in satisfying the 'optimality' criterion.

We are ready to introduce a 'simultaneous' algorithm which replaces our previous unsuccessful sequential optimization and control algorithm.

5.2.3. The 'S' Algorithm

Let $\mathbf{u}_0$ be the initial design. For the initial configuration we construct a near optimal control $\hat{q}(\epsilon_0)$ by selecting instantly optimal controls $q_j(x, t)$ on time subintervals $[t_i, t_{i+1}] \subset [0, T]$, $t_{i+1} - t_i < \epsilon_0$, where ϵ_0 is an *a priori* chosen number,

$\epsilon_0 > 2(E/\rho)^2 \bar{l}$, the constants E, ρ, $\bar{l}$ having the same meaning as in remark on controllability. Using $\hat{q}(\epsilon_0)$ constructed as in [36], section 2.19, chapter 2, we arrive at some number $\hat{C}$. We regard $\hat{C}$ as our upper bound on energy and optimize the design $\mathbf{u} \in \mathbf{U}$, changing $\mathbf{u}_0$ to $\mathbf{u}_1$ (A modified gradient method would provide a reliable design optimization algorithm for this step). We keep the constraint $\int_0^T \dot{W}_H \hat{q}_1(\epsilon_0) = \hat{C}$ active during this design optimization step. As we shall prove this constraint prevents the initiation of an infinite loop and forces a convergence to a pair $\{\hat{u}, \hat{q}\}$.

THE S-ALGORITHM. *Quality functional is* $J(u)$.

1. *Choose* ϵ_0, *choose* u_0.
2. *Compute* $q_{1,i}(\|q_{1,i}\| \leq 1)$ *on each interval* $[t_i, t_{i+1}]$ *using the instant optimization algorithm*: $\bar{q}_{1,i} = \delta(x - \xi_i)$, $\{\xi_i \in (0, l)|$
   $|\dot{W}(\xi_i, t_i)| \geq |\dot{W}(x, t_i)|, x \in (0, l)$,
   $q_{1,i} = \bar{q}_{1,i} \cdot \operatorname{sign}(\dot{W}(\xi_i, t_i))$.
3. *Derive* $W(x, T)$ *and*
   $$\hat{C}_1 = \sum_{i=1}^{n} (\dot{W}(\xi_i, t_i) q_{1,i}(x, t_i)\, \Delta t_i$$
   $(\Delta t_i = t_{i+1} - t_i)$.
4. *Choose* η_1, η_2 *(small numbers)*.
5. *Compute* $\operatorname{grad} J(u)_{|u_0}$.
6. *Compute* $u_0 - \zeta \cdot \operatorname{grad} J(u) = u_0 - \Delta u = u_1$
   $(\Delta u = \operatorname{grad} J(u)) \delta u = \zeta \operatorname{grad} J(u)$
   with ζ *chosen so that* $u_0 - \zeta \operatorname{grad} J(u)$ *is an admissible design.*
7. *Recompute* W, $\dot{W}$, $q_{2,i} = \delta(x, t_i)$
8. *Find if* $\hat{C}_2 = \sum \dot{W}(\xi_i, t_i) q_{2,i} \cdot \Delta t_i \leq \hat{C}_1$
   if not, change ζ.
   if yes, check if $|\hat{C}_2 - \hat{C}_1| < \eta_1$ *or* $|J(u_1) - J(u_0)| < \eta_2$.
   if yes, print $\hat{C}$, u *and end.*
9. *if no, go back to step* 5.

3.2.4. Justification of The S-Algorithm

Does this algorithm produce optimal design? The answer is no. In our previously given counter-example the design does not improve after the first step of the iterative process.

Let us prove that the numerical procedure outlined in the S-algorithm converges. That is, for any initial design $\mathbf{u}_0 \in U$ and any admissible initial control $q_0 \in Q$ there exists a pair $\{\bar{\mathbf{u}}, \bar{q}\}$ such that $\bar{\mathbf{u}} = \lim_{i \to \infty} \mathbf{u}_i$, $\bar{q} = \lim_{i \to \infty} q_i$, where the pair $\{\mathbf{u}_i, q_i\}$ is the intermediate design vector and control function after completion of the nth iterative step.

Proof: Let us compute the number

$$C_j = \left\{ \sum_{i=1}^{n} |W_j(\xi_i, t_i) q_{j,i}| \cdot \Delta t_i \right\}$$

at the end of jth iterative step, $j > 0$. The constraint assigned to C_j is $C_j \leqslant C_{j-i}$. Thus, the sequence C_j is monotone nonincreasing, sequence of positive numbers. It must converge to some positive number C_∞. Obviously $|q_{j,i} \cdot W_j| \neq 0$ on any positive time interval. The absolute value sign is inserted here for clarity, and is really unnecessary (see step 2 of the algorithm). Because of step 6, the numbers $J(\mathbf{u}_j)$ form a monotone non-increasing sequence of positive numbers. Because of step 8, the sequence will terminate when no further substantial improvement can be achieved in either the control, without worsening the design or in the design without worsening the control.

Thus, some 'compromise' optimal solution pair $\{\bar{u}, \bar{q}\}$ will be attained by this algorithm. This does not mean that this algorithm is entirely satisfactory. In fact, it is not suitable as a design algorithm for large, or complex mechanical or structural systems because of large number of computations required. Also it is not satisfactory from a purely theoretical point of view. The importance of optimal design and of extremal control have not been appropriately weighed. Thus, the process converges to a pair $(\bar{\mathbf{u}}, \bar{q})$, but does $(\bar{\mathbf{u}}, \bar{q})$ itself satisfy some optimality criterion?

5.2.5 Criticism of The S-Algorithm

Imposing the inequality $C < \hat{C}$ in each successive iteration we avoid the oscillation or the infinite cycle displayed in our previous counter-example. However, this algorithm (which resembles some commonly used numerical iterative schemes) is very inefficient. The numerical difficulty is caused by computation of the state of the system $W(x, t)$, of the derivatives $\dot{W}$, in evaluating W_H, $\dot{W}_H$ and of assessing if the control and the new design are admissible. For a large, or very complex system the number of computations becomes excessive, because no advantage is taken of any theoretical properties of either the optimal control process, or of the optimization. For example, ellipticity of the 'elasticity' operator, or symmetric properties of the 'control' operator were not utilized at all in our version of this algorithm.

The main computational difficulty in evaluating steps $\zeta \Delta u$ and Δt_2, in the S-algorithm consists in recomputing $\dot{W}(\xi_i, t_i)$ at each step after a change in the design $\mathbf{u}$. While this is more or less, a standard computational procedure, such lengthy recomputing at each step can be avoided with the use of an adjoint operator technique. This approach to computation of sensitivity was pioneered in articles [43,44].

Here, let us first outline the main idea of the adjoint operator numerical schemes pointing out some obvious computational advantages.

We note that the idea of using the adjoint operator equations in optimization theory is certainly not new and even the specific idea of the 'HK' trick was outlined in 1977 in [42]Without this simplifying step, or some similar numerical simplification, it is very difficult to handle numerically the optimization and design of complex large structural or mechanical systems, even when granted the availability of the most modern computer facilities.

The numerical procedure suggested by the S-algorithm can certainly be improved by incorporating the 'adjoint approach' in the steps 3, 7 and 8 of the algorithm.

5.2.6 'The Adjoint Operator 'Trick'

Let the state of the system be modelled by the equations

$$\left.\begin{aligned} Lz &= q(\mathbf{x}), \quad \mathbf{x} \in \Omega, L = L(\mathbf{u}) \\ Bz &= \mu(\mathbf{x}), \quad \mathbf{x} \in \partial\Omega, \end{aligned}\right\} \qquad 5.2.1$$

with constraints of the type

$$\psi_\alpha = \int_\Omega g_\alpha(\mathbf{x}, \mathbf{u}(x), z(x))\,\mathrm{d}\mathbf{x} + \int_{\partial\Omega} h_\alpha(x, u, z)\,\mathrm{d}\mathbf{x} \leq 0, \quad \alpha = 1, 2, \ldots, r. \qquad 5.2.1\mathrm{a}$$

In the rather straightforward, or you may say naive approach to optimization of a cost functional $J(z(\mathbf{u}), u(\mathbf{x}), \mathbf{x})$ with equations of state (2.1) and 'genuine' constraints (2.1a) all regarded as constraint conditions, we can compute the sensitivity of the functional J with respect to the design vector $u \in U$. Assuming Fréchet differentiability we compute the sensitivity of the cost functional:

$$\frac{\mathrm{d}\hat{J}}{\mathrm{d}\mathbf{u}} = \frac{\partial \hat{J}}{\partial \mathbf{u}} + \left\langle \frac{\partial \hat{J}}{\partial z} \cdot \frac{\partial z}{\partial \mathbf{u}} \right\rangle,$$

where

$$\hat{J} = J + \sum_{i=1}^{r} \nu^i \psi_i \;,$$

and ν^i are Lagrangian multipliers.

Now numerical methods can take over, iteratively improving the value of $J(\mathbf{u})$. This is the approach pursued in numerous engineering papers and tacitly assumed in the S-algorithm.

The main computational difficulty is caused by the term $\langle \partial\hat{J}/\partial z \cdot \partial z/\partial \mathbf{u} \rangle$. To evaluate this term we need to compute at each step of the iteration the state $z(\mathbf{u}(\mathbf{x}), \mathbf{x})$, or in a dynamic problem $z(\mathbf{u}(\mathbf{x}), \mathbf{x}, t)$.

The ideas proposed in [43,44] and subsequent papers permit to evaluate directly the sensitivity of the system (i.e. $\partial J/\partial \mathbf{u}$) without specifically evaluating $\partial z/\partial \mathbf{u}$. Numerically, this is of great importance.

Let us consider a variational form of (2.1), that is a bilinear functional

$$a(z, \mathbf{u}, \Lambda) = \langle Lz, \Lambda\rangle_\Omega - \langle q, \Lambda\rangle_\Omega + \langle Bz, \hat{\Lambda}\rangle_{\partial\Omega} - \langle \mu, \hat{\Lambda}\rangle_{\partial\Omega} + \sum_{i=1}^{r} \nu^i \psi_i(z, \mathbf{u}). \tag{5.2.2}$$

The equation $a(z, \mathbf{u}, \Lambda) = 0$ (2.2a) replaces (2.1) and (2.1a).

The products $\langle\ ,\ \rangle_\Omega$ and $\langle\ ,\ \rangle_{\partial\Omega}$ are products over Sobolëv spaces that are defined by integrals over the domain Ω and $\partial\Omega$, respectively.

For the sake of simplicity, let us assume that the loads $q(\mathbf{x})$ and $\mu(\mathbf{x})$ are independent of the design. Of course, this is not true in the combined design and control optimization problems. However, we try to avoid complications in explaining the basic idea of the Haug–Komkov (HK) 'trick'.

The adjoint variable Λ does not have to depend on the design. Therefore

$$\frac{\mathrm{d}}{\mathrm{d}\mathbf{u}}\langle Lz, \Lambda\rangle_\Omega = \frac{\mathrm{d}}{\mathrm{d}\mathbf{u}}\langle L^*\Lambda, z\rangle_\Omega = \left\langle \left(\frac{\mathrm{d}L^*}{\mathrm{d}\mathbf{u}}\right)\Lambda, z\right\rangle_\Omega + \left\langle L^*\Lambda, \frac{\mathrm{d}z}{\mathrm{d}\mathbf{u}}\right\rangle_\Omega. \tag{5.2.3}$$

Similarly

$$\frac{\mathrm{d}}{\mathrm{d}\mathbf{u}}\langle Bz, \hat{\Lambda}\rangle_{\partial\Omega} = \left\langle \left(\frac{\mathrm{d}B^*}{\mathrm{d}u}\right)\hat{\Lambda}, z\right\rangle_{\partial\Omega} + \left\langle B^*\hat{\Lambda}, \frac{\mathrm{d}z}{\mathrm{d}\mathbf{u}}\right\rangle_{\partial\Omega}.$$

Some explanations are necessary.

If $\mathbf{u}$ is an element of a Banach space $\mathcal{B}$, z is an element of a Sobolëv space H_1, Λ an element of H_2, L a map from H_1 to H_2, L^* from H_2 to H_1, $\mathrm{d}/\mathrm{d}\mathbf{u}$ is a map from $\mathbf{R}$ into $\mathcal{B}^*$, $\mathrm{d}L^*/\mathrm{d}\mathbf{u}$ is a tensor product and $\langle(\mathrm{d}L^*/\mathrm{d}\mathbf{u})\Lambda, z\rangle$ is not an inner product but a bilinear product whose values are in $\mathcal{B}^*$. Thus, $(\mathrm{d}L^*/\mathrm{d}\mathbf{u})\Lambda$ is a map from H_1 into $\mathcal{B}^*$, while $\mathrm{d}L^*/\mathrm{d}\mathbf{u}$ is a map from H_2 into a space of linear maps from H_1 into the space $\mathcal{B}^*$.

We use the notation $\langle\ ,\ \rangle$ rather sloppily. But the formal manipulation

$$\frac{\mathrm{d}}{\mathrm{d}\mathbf{u}}\langle f, q\rangle = \left\langle \frac{\mathrm{d}f(u)}{\mathrm{d}\mathbf{u}}, q\right\rangle$$

can be easily unravelled.

The operator $\mathrm{d}L^*/\mathrm{d}\mathbf{u}$ is easily computed following some formal rules, and so is the derivative $\partial\hat{J}/\partial\mathbf{u}$. Let the dual variable Λ obey the differential equation

$$L^*\Lambda = \sum \nu^i \frac{\partial\psi_\alpha}{\partial z} + \frac{\partial J}{\partial z}. \tag{5.2.4}$$

We recall that

$$\frac{\mathrm{d}\hat{J}}{\mathrm{d}\mathbf{u}}=\frac{\partial J}{\partial \mathbf{u}}+\sum_{\alpha=1}^{r} \nu^{i} \frac{\mathrm{d}\psi_{\alpha}}{\mathrm{d}\mathbf{u}}+\frac{\partial J}{\partial z} \frac{\partial z}{\partial \mathbf{u}},$$

or

$$\frac{\mathrm{d}\hat{J}}{\mathrm{d}\mathbf{u}}=\frac{\partial J}{\partial \mathbf{u}}+\sum_{i=1}^{r} \nu^{i} \frac{\partial\psi_{\alpha}}{\partial \mathbf{u}}+\frac{\partial J}{\partial z} \frac{\partial z}{\partial \mathbf{u}}+\sum_{i=1}^{r} \nu^{i} \frac{\partial\psi_{\alpha}}{\partial z} \frac{\partial z}{\partial \mathbf{u}}. \tag{5.2.5}$$

We consider the expression

$$\frac{\mathrm{d}}{\mathrm{d}\mathbf{u}}\langle L^{*}\Lambda, z\rangle=0. \tag{5.2.6}$$

This equality is true because $q(x)$ is independent of the design, and so is $\Lambda(x)$. But

$$\langle L^{*}\Lambda, z\rangle=\langle\Lambda, Lz\rangle=\langle\Lambda, q\rangle.$$

Therefore,

$$\frac{\mathrm{d}}{\mathrm{d}\mathbf{u}}\langle L^{*}\Lambda, z\rangle=\left\langle\frac{\mathrm{d}L^{*}}{\mathrm{d}\mathbf{u}}\cdot\Lambda, z\right\rangle+\left\langle L^{*}\Lambda, \frac{\partial z}{\partial \mathbf{u}}\right\rangle=0. \tag{5.2.7}$$

Thus,

$$\left\langle L^{*}\Lambda, \frac{\partial z}{\partial \mathbf{u}}\right\rangle=-\left\langle\frac{\mathrm{d}L^{*}}{\mathrm{d}\mathbf{u}}\Lambda, z\right\rangle$$

and

$$\frac{\mathrm{d}\hat{J}}{\mathrm{d}\mathbf{u}}=\frac{\partial\hat{J}}{\partial \mathbf{u}}+\left\langle L^{*}\Lambda, \frac{\partial z}{\partial \mathbf{u}}\right\rangle=\frac{\partial\hat{J}}{\partial \mathbf{u}}-\left\langle\frac{\partial L^{*}}{\partial \mathbf{u}}\Lambda, z\right\rangle. \tag{5.2.8}$$

This is the basic idea of the '(HK) trick'.

Formula (2.8) allows us to compute the sensitivity and the gradient of the cost functional without computing at each state of the numerical process the state sensitivity tensor $\partial z/\partial\mathbf{u}$.

5.2.7. Other Possible Numerical Approaches

One may look at the problem of simultaneous control and design problem as a typical multicriterion optimization problem. From this point of view several of the difficulties encountered in our previous discussion are easily comprehended and, in fact, are to be expected. Thus, we can deal with it by invoking one of the known weighing, penalty or compromise techniques. One could, for example, regard this from the cooperative criteria point of view. The control and optimization criteria are regarded as simultaneous controls and one of the (known) possible compromise solutions may be applied. The best known are the Pareto and Nash equilibria. The Nash approach (that is the dog-eats-dog criterion) is not suitable for this type of problem. The Pareto, or cooperative equilibrium approach appears to be best suited.

The subject of cooperative equilibria can be literally transplanted from the mathematical economics. An extensive research into applications of Pareto equilibrium principle to some mechanical engineering design problems has been carried out at the University of California at Berkeley by G. Leitmann and his colleagues. Even a partial review of this approach or of the competing, Nash equilibrium theory is outside the scope of this monograph. While we by-pass the discussion of important developments in applications of the Pareto approach some comments can be made concerning its applications to engineering optimization.

The Pareto process can be regarded as a generalization of a simple control process where only one pay-off functional is considered. A well-known approach to Pareto optimization consists in imbedding the problem in an n-dimensional Euclidean space, whose orthogonal basis consist of vectors

$$\xi_i = \frac{\mathrm{d}J_i(x(t))}{\mathrm{d}\mathbf{x}}.$$

The gradient of J_i vanishes at the extremum (hopefully a maximum) of $J_i(x)$. However, this condition assigned to time $t = T$ prevents optimization of J_i, $j \neq i$, at $t = T$.

A compromise solution adopted by Pareto and subsequently by other writers consists in formulating adjoint functional $J = \sum_{i=1}^{n} c_i J_i$, where c_i is the weight assigned to the i-th player reflecting how much importance his wishes carry compared to the other players. We normalize c_i, by requiring $c_i > 0$, for $i = 1, 2, \ldots n$, and $\sum_{i=1}^{n} c_i = 1$. Then the theoretic approach consists of looking for a final state $\mathbf{x}(T)$ such that

$$\mathrm{d}J|_{t=T} = \langle \xi, \mathrm{d}\mathbf{x} \rangle|_{t=T} = 0.$$

This is, of course, only a necessary condition for optimality, and signs of appropriate functionals of second derivatives have to be examined to ascertain the maximality of J.

Chapter 5

References for Chapter 5

[1] V.G. Litvinov, Optimal control of coefficients in elliptical systems, publication of Ukrainian Academy of Sciences, Institute of Mathematics, Kiev, 1979.

[2] W. Prager, Introduction to structural optimization, Springer Verlag, Berlin, 1974.

[3] S. Prager and W. Prager, A note on optimal design of columns, Int. J. Mech. Sci., Vol. 21, 1979, p. 249-251.

[4] V. Komkov and N. Coleman, "Optimality of Design and Sensitivity Analysis of Beam Theory," Int. J. Control, Vol. 18, No. 4, 1973, pp. 731-740.

[5] E.J. Haug, "Two Methods of Optimal Structural Design," Developments in Mechanics, Vol. 5 (Weiss, H.J., Young, D.F., Riley, W.F., and Rogge, T.R.,Ed.), Iowa State University Press, 1969, pp. 847-860.

[6] E.J. Haug, "A Gradient Projection Method for Structural Optimization," Optimization of Distributed Parameter Structures (E.J. Haug and J. Cea,Ed.), Sijthoff & Noordhoff, Alphen aan den Rijn, Netherlands, 1981, pp. 446-473.

[7] M.M. Vaǐnberg, Variational methods for investigation of nonlinear operators, Holden Day, San Francisco, 1963 (Translated from Russian).

[8] Z. Nashed,"Differentiability and related properties of non-linear operators: Some aspects of the role of differentials in non-linear functional analysis", in L.B. Rall, ed., Non-

linear functional analysis & applications, Academic Press, New York, 1971, pp. 103-309.

[9] T. Kato, Perturbation Theory for Linear Operators, Springer Verlag, Berlin and New York, Die Grundlehren der mathematischen Wissenschaften, vol. 132, (1966).

[10] M. Frechét, La notion de differentielle dans l'analyse generale. Ann. Soc. de l'Ecole Norm. Super., 42, (1925), p. 293-323.

[11] R. Gateaux, Sur les functionelles continues et les functionelle analtiques, Comptes Rendues, 157, (1913), p. 325-327.

[12] N. Aronszajn, Approximation methods for eigenvalues of completely continuous symmetric operators, in Proc. Symp. Spectral Theory and Differential Problems, Oklahoma State Univ., Stillwater, Okla, 1951.

[13] V, Komkov, The critical point theory and the variational principles, Ars. Polona, Scientific Papers of the Institute of Civil Engineering, Technical University of Wrocław, Monograph #30, 1985.

[14] J.L. Lagrange, Sur le figure de la colonnes, Miscelenea Taurinensia, Vol. V, reprinted in 1970 (p. 123-125)

[15] J. Hadamard, Lectures on the Calculus Variations(in French). Gauthier-Villards, Paris(1910).

[16] O. Pironneau, Optimal shape design for elliptic systems, Springer Verlag, Berlin 1984.

[17] V. Komkov, A dual form of Noether's theorem with applications to continuum mechanics, J. Math. Anal. Appl. 75, #1, 1980, p. 251-269.

[18] V. Komkov, The optimization of the domain problem, I. Basic Concepts, J. Math. Anal. Applic., vol. 82, #2, 1981, p. 317-333.

[19] S. Bergman and M. Schiffer, Kernel functions and elliptic differential equations of mathematical physics, Academic Press, New York, 1953.

[20] P. Garabedian, Schwarz's lemma and the Szegö kernel function. Trans. Amer. Math. Soc. 67,(1949), 1-35.

[21] J. Hadamard, Memoire sur probleme d'analyse relatif a l'equilibre des plaques elastiques encastrées, Deuvres de Jacques Hadamard, C.N.R.S., Paris, 1968.

[22] Kyung K. Choi, Shape design sensitivity analysis of displacement and stress constraints, J. Structural Mech. 13,(1985), no. 1. 27-41.

[23] Zolesio,J.P., "Identifiçation de domains par Deformations." Theѐse d'État, Université de Nice, 1979.

[24] Zolesio, J.P., "Gradient des coûts Governés par des problèmes de Neumann posés sur des Ouverts Anguleux en Optimisation de Domain," CRMA-Rep. 1116. University of Montreal, Canada, 1982.

[25] Lord Rayleigh, The theory of sound, London 1894, reprinted by Dover Press, N.Y, 1961.

[26] J.P. Zolesio, The material derivative (or speed) method for shape optimization, in Optimization of Distributed Parameter Structures(E.J. Haug and J. Cea, eds.), Sijthoff & Noordhoff, Alphen aan den Rijn, Netherlands, 1981, pp. 1089-1151.

[27] R. Courant and D. Hilbert, Methods of Mathematical Physics, Vol. I, II, Interscience Press, New York, 1962.

[28] N.V. Banichuk, Shape optimal design, Plenum Press, New York, 1985.

[29] O.C. Zienkiewicz and Campbell, J.S.: 'Shape Optimization and Sequential Linear Programming' in R.H. Gallagher and O. C. Zienkiewicz (eds.), Optimum Structural Design, Wiley, New York, 1973, pp. 109-126.

[30] V. Bourbaki, Espaces vectoriels topologiques, Chapter II, (see exercises), Volume V.

[31] V. Komkov , Control Theory, Variational Principles and Optimal Design of Elastic Systems, Proceedings of International Conference in Norman, Oklahoma, March 1977, Wiley, New York, 1978.

[32] V. Komkov, and Coleman, N., An Analytic Approach to some Problems of Optimal Engineering Design , Archives of Mechanics 27, (1975), 565-575.

[33] A.L. Dontchev, Perturbations, Approximations and Sensitivity Analysis of Optimal Control Systems, Lecture Notes in Control and Information Science No. 52, 1983, Springer-Verlag, New York, Berlin.

[34] A.L. Dontchev, and Veliov, V.M., 'Singular Perturbation in Mayer's Problem for Linear Systems', SIAM J. Control Optim. 21,(1983), 566-581.

[35] V. Komkov, 'Sensitivity Analysis in Some Engineering Applications', in V. Komkov(ed.) Sensitivity of Functionals with Applications to Engineering Sciences, Lecture Notes in Mathematics Series, No. 1086, Springer-Verlag, Berlin, Heidelberg, New York, 1984.

[36] V. Komkov, The Optimal Control Theory for the Damping of Vibrations of Simple Elastic Systems (in English). Lecture Notes in Mathematics, No. 253, Springer-Verlag, Berlin, 1972, Russian translation, 1975.

[37] W. Prager, 'Optimal Design of a Statically Determinate Beam for a Given Deflection', J. Mech. Sci. 13(1971), 893.

[38] N.V. Banichuk, 'Optimality Conditions and Analytical Methods of Shape Optimization' in E.J. Haug and J. Cea(eds.), Optimization of Distributed Parameter Structures, Sijthoff & Noordhoff, Alphen aan den Rijn, Netherlands, 1981, pp. 973-1004.

[39] N.V. Banichuk, 'Design of Plates for Minimum Deflection and Stress', in E.J. Haug and J. Cea,(eds.), Optimization of Distributed Parameter Structures, Sijthoff & Noordhoff, Alphen aan de Rijn, Netherlands, 1981, pp. 333-361.

[40] E.J. Haug, and Arora, J. S., 'Distributed Parameter Structural Optimization for Dynamic Response' in E.J. Haug and J. Cea(eds.) Optimization of Distributed Parameter Structures Sijthoff & Noordhoff, Alphen aan de Rijn, Netherlands, 1980.

[41] E.J. Haug, 'A Review of Distributed Parameter Structural Optimization Literature', in E.J. Haug and J. Cea(eds.) Optimization of Distributed Parameter Structures, NATO Advanced Study Series, Applied Sciences #49, Sijthoff and Noordhoff, Alphen aan de Rijn, Netherlands 1981, pp. 3-81.

[42] E.J. Haug and Komkov, V., 'Sensitity Analysis in Distributed Parameter System Optimization, J. Optimization Theory Appl. 23(1977), 445-463.

[43] C.L. Irwin and Komkov, V., 'Sensitivity Analysis and Model Optimization for Reaction-Diffusion Systems', DOE Report, 1984.

[44] E.J. Haug, K.C. Choi, and V. Komkov, Sensitivity Theory for Structural and Mechanical Systems, Academic Press, New York, New York, 1986.

[45] C.L. Irwin and V. Komkov, Sensitivity Analysis and Model Optimization for Reaction-Diffusion Systems, Journal of Optimization Theory and Applications: Vol. 44, No. 4, p. 569-584.

[46] V. Komkov and V. Dannon, Design Optimization for Structural or Mechanical Systems Against the Worst Possible Loads, Acta Appalic. Mathematicae, 8 (1987), p. 37-64.

[47] J.P. Zolesio, Semi-Derivatives of Repeated Eigenvalues, Optimization of Distributed Parameter Structures, Edited by E.J. Haug and J. Cea, Sijthoff and Noordhoff, Alphen aan den Rijn, Holland, pp. 1475-1491, 1981.

Appendix to Chapter 5

Bibliographical Note

We have literal explosion of knowledge concerning the principles of optimal or improved design of engineering structures. List of relevant articles and related bibliographies have been published at regular intervals. A rough estimate is of between 10,000 and 20,000 of original articles, papers and reports.

Best known western literature reviews are those of A.M. Brandt ([1], 1968), W.R. Micks ([2], 1958), E.J. Haug ([3], 1981).

Another review is now overdue. The Soviet list of sources include T.N. Driyaeva([4], 1968) containing about 1500 items, V.I. Mazalov and Iu. V. Nemirovskii([5], 1975), M.I. Reitman and G.S. Shapiro([6], 1978). Books that contain extensive bibliography include N.V. Banichuk's ([7], 1983), K.A. Lurie([11], 1975), R.H. Gallagher and O.H. Zienkiewicz([8], 1973), E.J. Haug([10], 1969), E.J. Haug and J. Arora([9], 1975).

This note refers strictly to the mechanical and structural optimization bibliographic reports. Even articles dealing with closely related topics, such as optimization of electrical and electronic networks, optimization of chemical of bio-chemical processes have been deliberately ignored.

A sample of related mathematical literature dealing with the applications of modern control theory in general, and of Pontryagins maximality principle in particular may be found in the references of the book of Grinev and Filippov[12]. Also see Haug's Report [13], and article of Lurie [14].

The problems concerning optimal use of different materials (including reinforced concrete and reinforced plastics have been largely ignored in this volume.

For a sample of articles and related bibliography see, for example, Krakovskiĭ et. al., [15], Baĭkov and Skladnev [16], Alspaugh and Huang [17], or the monograph of Banichuk [7].

References for the Appendix

[1] A.M. Brandt, Zestawienie bibliograficzne prac dotyczacych optimalizacji ksztaltow konstrukcji. In: Metody Optimalizacji ustrojów odkształcalnych, 1968.

[2] W.R. Micks, Bibliography of literature on optimum design of structures and related topics. Rand Corp. Rsch. Memo RM-2304, DDC NAD 215771, Dec. 1958.

[3] E.J. Haug, A review of distributed parameter structural optimization literature, in "Optimization of Distributed Parameter Structures", E.J. Haug and J. Cea, editors, Vol. I., Sijthoff and Noordhoff Publ. Co., Alphen an den Rijn, Netherlands, 1981, pp. 3-74.

[4] T.N. Driyaeva, Optimal design of structures, Bibliography and Index, (in Russian) ONTI, TS.,N.I.I.E., Pselstroi, 1968.

[5] V.I. Mazalov and Iu. V. Nemirovskii, Optimal Design of Structures, Bibliography and Index for the period 1948-1974, Part I. Novosibirsk, Institute for Hydrodynamics of Sibirian Branch of Acad. Sc. U.S.S.R., 1975.

[6] M.I. Reitman and G.S. Shapiro, Optimal design of deformable solid bodies of deformable solid bodies, in: Mechanics of Deformable solid bodies, #12, 1978, pp.5-90.

[7] N.V. Banichuk, Shape optimal design, Plenum Press, N.Y., 1985. (Translated by V. Komkov).

[8] R.H. Gallagher and O.C. Zienkiewicz, editors, Optimum Structural Design, John Wiley & Sons, London, 1973.

[9] E.J. Haug, J.S. Arora, Distributed Parameter Structural Optimization for Dynamic Response in E.J. Haug and J. Cea (eds.) Optimization of Distributed Parameter Structures Sijthoff & Noordhoff, Alphen aan de Rijn, Netherlands, 1980.

[10] E.J. Haug, Optimal Design of Structural Elements, Lecture Notes, Dept. of Mech. and Hydr., Univ. of Iowa, 1969.

[11] K.A. Lurie, Optimal control in problems of mathematical physics, Nauka, Moscow, 1975.

[12] V.B. Grinev and A.P. Filippov, Optimization of structural elements with respect to mechanical properties, Nauk. Dumka, Kiev, 1975.

[13] E.J. Haug, U.S. Army Material Command Report, 1973.

[14] A.I. Lurie, Applications of the maximality principle to some simple problems of mechanics, Trudy Leningrad. Pol. Inst. #252, 1965.

[15] M.B. Krakovskiĭ, G.K. Haidukov and A.V. Shapiro, Optimal design of reinforced concrete shells for roofs of buildings, Trudy (of the 10th all Union conference on the theory of shells and plates), Kutaisi, 1975, Metsnereba Publ. Tbilisi, 1975, p. 593-601.

[16] V.N. Baikov and N.N. Skladnev, Application of a probabilistic approach to the optimization of reinforced concrete beams. Sbornik Trudov Mosc. Univ. (Engineering Institute), #151, 1977, p. 3-10.

[17] D.W. Alspaugh and S.N. Huang, Minimum weight design of axisymmetric sandwich plates, AIAA Journal, 14, No. 12, 1976, p. 1683-1689.

Chapter 6

Some Comments on Properties of the Sensitivity Function

6.1 The state sensitivity function.

Let the state equations of a mechanical system be of the form $\phi_i(\underset{\sim}{x},t)\ w, w_x, w_t) = 0$, $i = 1,2,\ldots m$, $x \in R_n$, $t \in \mathbb{R}_+$, $w: \{(\Omega \subset \mathbb{R}^n) \times \mathbb{R}_+)\} \to \mathbb{R}$. w_x, w_t are defined in the weak (Sobolev sense). w is called the state, t will be identified with the time, $\underset{\sim}{x}$ with generalized "position" coordinates. The state $w(t,\underset{\sim}{x})$ is a vector in a normed space H_1. Some holonomic constraints are assigned to the system $\psi_j(t,x,w) \leq 0$, or $\psi_x(t,\underset{\sim}{x},w) = 0$ $\psi_\alpha: \{\mathbb{R}_+ \times \Omega \times H_1\} \to \mathbb{R}$. The spatial domain $\Omega \subset \mathbb{R}^n$ is a manifold, called the (admissible) kinematic manifold. The state function w depends on a number of physical parameters $h_i(\underset{\sim}{x})$, $i = 1, 2\ldots k$. The vector $\underset{\sim}{h}(\underset{\sim}{x})$ is an element of a Banach space B, that could be L_1 or L_∞. $\underset{\sim}{h}(\underset{\sim}{x})$ belongs to $U \subset B$ called the admissible design. To define the sensitivity function w_n we first define the Gateaux difference in the direction of vector

$\underset{\sim}{\eta} \in B$: $\Delta_{\varepsilon\underset{\sim}{\eta}} w(\underset{\sim}{h}_0) = w(\underset{\sim}{x},t,\underset{\sim}{h}_0(\underset{\sim}{x}) + \varepsilon\underset{\sim}{\eta}(\underset{\sim}{x}))$ - $w(x,t,h_0(x))$ - $w(\underset{\sim}{x},t,\underset{\sim}{h}_0(\underset{\sim}{x}))$ where ε is a real number.

Let ϕ be a mapping from B into H_1, let <,> denote a bilinear product $\langle f,g\rangle \in H_1$, $f \in B$.

If $\Delta_{\varepsilon\underset{\sim}{\eta}} w(\underset{\sim}{h}_0)$ can be represented in the form

$$\Delta_{\varepsilon\underset{\sim}{\eta}} w(\underset{\sim}{h}_0) = \varepsilon < \phi\ (t,\underset{\sim}{x},\underset{\sim}{h}),\ \underset{\sim}{\eta} > + \zeta\ \varepsilon\underset{\sim}{\eta})$$

such that $\frac{\|\zeta\,(\varepsilon\eta)\|}{\|\varepsilon\eta\|} \to 0$ as $\varepsilon \to 0$ and $\phi(t,\underset{\sim}{x},\underset{\sim}{h})$ is independent of η then we shall call $\phi(t,\underset{\sim}{x},\underset{\sim}{h})$ the Fréchet derivative of w with respect to $\underset{\sim}{h}$ and denote $\phi = w_h$. (See [1] and [2].) For small values of ε we designate the function $w_{\underset{\sim}{h}}$ the "sensitivity" function for the state of the system. Roughly speaking, the sensitivity of the state w(x,t) with respect to the design vector h reflect how small changes in h affect the state w(x,t). The sensitivity w_h is an abstract derivative of w with respect to h. If the value of this derivative is large this means that small changes in the design may cause large changes in the state of the system.

Similarly one may define the sensitivity of other functions, or functionals such as the cost functional in a variational problems with respect to the design.

In this chapter we do not attempt a comprehensive, or even partial coverage of the sensitivity of the state problem. For an extensive discussion of the sensitivity of structural systems see the monograph of Haug, Choi and Komkov[3].

2. The sensitivity equation

We introduce some elementary lemmas permitting us to manipulate the properties of the sensitivity function $w_h : B \to H_1$. Strong or weak continuity and differentiability of w_h with respect to the variables $\underset{\sim}{x}$,t is defined in the usual manner.

Lemma 1:

If continuous differentiability is assumed for w and for h(x) and w_h exists then

$$\frac{\partial u_h}{\partial x} = (w_x)_{\underset{\sim}{h}}.$$

Theorem 1:

If sufficient smoothness assumptions are made then the sensitivity function for the system ϕ and the constraints χ

$$(1^a) \quad \phi(w, w_x, w_t, \underset{\sim}{x}, t, h(x)) = 0,$$

$$(1^b) \quad \chi_i(t, \underset{\sim}{x}, w) \leq 0,$$

obeys a quasilinear equation:

$$(2) \quad a_2 (w_h)_t + \sum_{i=1}^{n} a_{1_i} \cdot (w_h)_{x_i} + a_o \cdot w_{\underset{\sim}{h}} + c = 0$$

Proof. Assuming that all derivatives, that are displayed here, exist, we use the chain rule

$$\frac{d\phi}{dh} = \frac{\partial\phi}{\partial h} + \frac{\partial\phi}{\partial u} u_h + \frac{\partial\phi}{\partial u_x}(u_h)_x + \frac{\partial\phi}{\partial u_t}(u_h)_t + \sum_i (\lambda_i \frac{\partial\chi_i}{\partial u}) u_h = 0,$$

or

$$(2^a) \quad c_o(\underset{\sim}{\lambda}, t, x, u, u_x, u_t h) + a_o(\ldots) u_h + \underset{\sim}{a}_1(\ldots) (u_h)_{\underset{\sim}{x}} + a_2(\ldots)(u_h)_t = 0.$$

We note that $\phi(x, t, w, w_x, w_t, \underset{\sim}{h}) = 0$ for an $h \in U \subset B$, that is for any admissible value of the design variable $\underset{\sim}{h}(\underset{\sim}{x})$ $\cdot(\lambda_i \geq 0$ are Lagrangian multipliers). U is the set of admissible designs.

Lemma 3. If L is a linear differential operator $L_{\underset{\sim}{h}}$ is defined in the neighborhood of $h_o \in R^m$ then

$(Lu)_h = L_{\underset{\sim}{h}}\ u + L_{\underset{\sim}{h}}$.

Theorem 2:

If $\phi(\ldots) = 0$ is a linear partial differential equation, then the sensitivity equation (2.1) is again a partial differential equation of the same form.

Outline of the proof:

Basically the proof follows from the formula 2.1, by putting $\phi(u,u_x,u_t,x,t,h) = L(h)\underset{\sim}{u} = q(\underset{\sim}{x},t)$, where we may consider $\underset{\sim}{u}$ to be a vector rather than scalar element of an appropriate space H. For example, if second derivatives $u_{x_i x_j}$ occur in the state equation then we replace $\underset{\sim}{u}$ by the vector

$$(\underset{\sim}{u}, v_1, v_2 \ldots v_n) = \{u, v\} = u_{x_1}, \ldots, v_n = u_{x_n},$$

$$\text{or by} \begin{bmatrix} u \\ v_i \end{bmatrix} = \underset{\sim}{u} . \qquad \text{Thus } u_{x_i x_j} = u_{i_{x_j}} .$$

This is a standard method of conversion of a high order system to a system of first order equations. Thus $\phi = 0$ is converted to a system of linear differential first order equations

$L(\underset{\sim}{h})\ u_i = q_i(\underset{\sim}{x},t)$, $\underset{\sim}{h} \in \kappa$, which is of the form

$$(3) \quad \sum_{i,j} a_{ij}\, U_{i,_{kj}} + \sum_k a_k\, U_{k,_t} + \sum_i b_i\, U_i + C = q.$$

Thus, the sensitivity equation is given by

$$(3^a) \quad \sum_{i,j} (a_{ij_{\underset{\sim}{h}}} U_{i_{x_j}} + a_{ij}\, U_{i_{\underset{\sim}{h},x_j}}) + \sum a_k\, U_{k_{\underset{\sim}{h},t}}$$

$$+ b_{\underset{\sim}{h}} U + bU_{\underset{\sim}{h}} + C_{\underset{\sim}{h}} = 0,$$

$$\text{or } \sum_{i,j} a_{ij} U_{i_{\underset{\sim}{h},x_j}} + \sum_k a_k U_{k_{\underset{\sim}{h},t}}$$

$$+ \sum_i \beta_i U_{h_i} + C_{\underset{\sim}{h}} = 0.$$

The constraints have been ignored. It should be clear, that if the Lagrangian multipliers are not functions of $\underset{\sim}{h}$ then the general term

$\sum_i \lambda_i \chi_{h_i}$ will be included in the equation ($3^{\underline{a}}$)

modifying only the terms $C_{\underset{\sim}{h}}$ and $\beta_i \cdot U_{h_i}$. This completes the outline of the proof.

<u>Comments.</u> If the terms $a_{ij}U_{i,x_j}$, $a_k U_{k,t}$ $b_i U_i$ and C are of the same order of magnitude, and the same may be said of all terms in ($3^{\underline{a}}$) then no serious problems arise in computing the sensitivity of the system or the sensitivity of a related functional $J(u,u_x,u,\underset{\sim}{x},t,\underset{\sim}{h}(x))$. In that case the sensitivity equation is a mirror image of the original state equation.

<u>Example.</u>

Consider the state function $\hat{u}(h(x),x,t) = u(x,t)$ and the linear equation $m(x,t)\,\ddot{u}(x,t) +$

$$\underset{\sim}{k} \cdot \text{grad } (u) = \underset{\sim}{q}(t), \quad (\cdot = \frac{\partial}{\partial t}).$$

The sensitivity equation is

$$m(x,t)\,\ddot{S}(x,t) + (k \text{ grad } S(x,t) = 0,$$

For a nonlinear system

$$m(x,t)\,\ddot{u}(x,t) + k(u \cdot \text{grad } u) = q(x,t)$$

the sensitivity equation is

$$m\,\ddot{s} + k(u \cdot \text{grad } s) + k(s \cdot \text{grad } u) = 0$$

<u>An example</u>: Burger's equation.

$$(4) \quad \phi(u,u_x,u_t,\varepsilon) = u_t + u\,u_x - \varepsilon\,u_{xx} = q(x,t)$$

$$u = u(\varepsilon).$$

and $4^{(H)}$ obtained by setting $q(x,t) \equiv 0$.

Then :

$$(5) \quad \frac{d\phi}{d\varepsilon} = \frac{\partial\phi}{\partial\varepsilon} + \frac{\partial\phi}{\partial u}u_\varepsilon + \frac{\partial\phi}{\partial u_x}u_{\varepsilon_x} + \frac{\partial\phi}{\partial u_{xx}}u_{\varepsilon_{xx}} + \frac{\partial\phi}{\partial u_t}u_{\varepsilon_t}$$

$$= -u_{xx} + u_x\,u_\varepsilon + u\,u_{\varepsilon_x} - \varepsilon\,u_{\varepsilon_{xx}} + u_{\varepsilon_t}$$

$$= -u_{xx} + u_x\,s + u\,s_x + s_t - \varepsilon\,s_{xx} = 0,$$

where $s(x,t,\varepsilon) = u_\varepsilon$.

Behavior of the sensitivity equation for large values of x can be deduced from our assumptions:

$$\begin{cases} \lim\limits_{x\to\pm\infty} u(x,t) = u_\infty = \text{constant}\,, \\ \lim\limits_{x\to\pm\infty} u_x = 0\,, \\ \lim\limits_{x\to\pm\infty} u_{xx} = 0\,. \end{cases}$$

It is convenient to use non-standard symbolism.

Thus if $x \in {}^*R_\infty$ then

$$\begin{cases} u(x,t) = u_\infty + \xi, \\ u_x(x,t) = \eta, \\ u_{xx}(x,t) = \zeta, \end{cases}$$

and s satisfies the equation

$$-\xi + \eta \cdot s + (u_\infty + \xi)\, s_x + s_t - \varepsilon\, s_{xx} = 0 \qquad (5^a)$$

where $\xi, \eta, \zeta \in \mu(o)$ (Monad of zero).

The original equation (4^H) has a conservation law derived from

$$(4^a) \quad \frac{\partial u}{\partial t} = \frac{\partial}{\partial x}\left(\varepsilon u_x - \frac{u^2}{2}\right)$$

If conditions at infinity are

$$\lim_{x \to \pm\infty} u\,(x,t) = \text{constant},$$

$$\lim_{x \to \pm\infty} u_x = 0,$$

then $$\frac{\partial}{\partial t} \int_{-\infty}^{+\infty} u(x,t)\, dx = [\varepsilon\, u_x - u^2/2\,]_{-\infty}^{+\infty} = 0$$

and

$$\int_{x=-\infty}^{x=+\infty} u(x,t)\, dx \text{ is an invariant.}$$

A similar conservation law can be derived for equation (4) if, for example,

$$q(x,t) = q(x) = \frac{\partial Q(x)}{\partial x},$$

with

$$\lim_{x \to \pm\infty} Q(x) = \text{constant}.$$

The invariants of the sensitivity equation can be derived from the observation that the equation(3) can be reqritten as

$$\frac{\partial}{\partial x}(+ u_x - (us) + \varepsilon \cdot s_x) = s_t.$$

Thus, if $\lim\sup_{x \to \pm\infty} \{|s| + |s_x|\} < \infty$,

and

$$[u_x - (u \cdot s) + \varepsilon\, s_x]_{-\infty}^{+\infty} = 0,$$

then

$$\int_{-\infty}^{+\infty} (s_t)\, dx = \frac{d}{dt} \int_{-\infty}^{+\infty} s\, dx = 0,$$

and $\int_{-\infty}^{+\infty} s(x,t)\, dx \equiv \text{constant}.$

Therefore, redesigning Burgers' flow by changing the parameter ε amounts to redistributing the sensitivity function $s(x,t)$ over the real line $(-\infty<x<+\infty)$ while keeping "the total amount of sensitivity" $(\int_{-\infty}^{+\infty} s\, dx)$ constant.

Differentiability of the state.

Let $a(u,v)$ be a bilinear form i.e. $a(u,v) = \langle L(h)\, u,\, v\rangle_{L^2(\Omega)}$ (according to the Lax-Milgram representation theorem). u is the state(or displacement) of the system. $\langle,\rangle$ is the usual $L^2(\Omega)$ product. The notation L(h) implies that the operator L depends on the design vector h.

If L(h) is a positive definite operator then L(h) can be written in the form $L(h) = A(h)\, A^*(h)$. Thus $a(u,v) = \langle A\, u,\, A\, v\rangle$ where A depends on a design vector $\tilde{h}(x)$. u and v are $L^2(\Omega)$ functions, but v can be selected from a set of admissible generalized forces v where v could be a dense subset of $L^2(\Omega)$.

Differentiability of a(u,v) with respect to the design $\tilde{h}$ implies that

$$(6) \quad < L(\tilde{h} + \varepsilon \tilde{\eta})\, u,\ v > = < L(\tilde{h})\, u,\ v > + < L\, u,\ v >_{\tilde{h}} + r(u,v,\varepsilon\tilde{\eta}),\ \varepsilon \in R,\ \|\tilde{\eta}\| \leq 1,$$

where

$$(6^a) \quad \frac{\| \zeta(u,v)}{\varepsilon \| \tilde{\eta} \|} \to 0 \text{ as } \varepsilon \to 0.$$

However, whether a(u,v) is Fréchet differentiable or not, we have

$$< L\ (\tilde{h}+\varepsilon\tilde{\eta})\ u,v > = < L\ (\tilde{h})\ u,v > + \varepsilon < \alpha\ (\tilde{\eta})\ L(\tilde{h})\ u,\ v > + \varepsilon^2 < \beta\ (\tilde{\eta})\ L(\tilde{h})\ u,v > + o(\varepsilon^3).$$

Kato[5] shows that a perturbation of a linear operator $L(\tilde{h})$ results in the following expansion

$$< L(\tilde{h}+\varepsilon\tilde{\eta})\ u,v> = < L\ (\tilde{h})\ u,\ v + < L\ u,v_{\tilde{h}}> +$$

$r\ (u,v,\varepsilon\tilde{\eta})$ where $r(u,v,\varepsilon\eta)$ may not have the property (6^a). However, if the remainder term $r(u,v,\varepsilon\tilde{\eta})$ is bounded by $\varepsilon^2\|\eta\| \cdot a(u,v)$ then the conclusion (6) and (6^a) follows.

Moreover, there exist operators $\alpha(\tilde{\eta})$, $\beta(\tilde{\eta})$ such that

$$< L(h+\varepsilon\eta)\ u,v > = a(u,v)_h + < L(h)\ u,v > + r =$$

$$< \{A(h) + \varepsilon[(\alpha \cdot A(h)) + \varepsilon\beta\ A(h)]\}u,\ A^*(h)\ v >.$$

The operator $A(\underset{\sim}{h}) \cdot (I + \varepsilon\alpha + \varepsilon^2\beta) =$

$A \cdot I_\varepsilon$, replaces $A(\underset{\sim}{h})$.

Here α, β are continuous operators from $L^2(\Omega)$ to $L^2(\Omega)$. Thus, if $\|\alpha\|$ and $\|\beta\|$ are bounded by some constant and ε is chosen sufficiently small, the operator I_ε is invertible.

It follows after some manipulation that the operator $[L(\underset{\sim}{h}+\varepsilon\underset{\sim}{\eta})]^{-1}$ is defined, and is given

by $A(\underset{\sim}{h})^{-1} I_\varepsilon^{-1} A(\underset{\sim}{h})^{-1}$.

(A similar line of argument can be found in the in the Haug-Choi-Komkov monograph [3], earlier, in the paper of Haug and Rousselet [4]). This implies that the Fréchet differentiability of the bilinear functional a(u,v) implies the Fréchet differentiability of the state $\underset{\sim}{u}$ with respect to the design $\underset{\sim}{h}(x)$, provided the load is either independent of the design, or is a smooth function of the design.

In the static case we have the following corrolaries.

Corrolary 1.

If the basic potential energy functional a(u,v) is a Fréchet differentiable function of the design $\underset{\sim}{h}(x)$, then the sensitivity of the state function $u_{\underset{\sim}{h}}$ is a continuous function of the design.

Corrolary 2.

A discontinuity in the sensitivity function $u_{\underset{\sim}{h}}(x)$ at the point $\underset{\sim}{h} = \underset{\sim}{h}_0(x)$ implies that the potential functional has at best only directional derivatives with respect to the design variable at the point $\underset{\sim}{h}(x) = \underset{\sim}{h}_0(x)$ in the admissible set of design space.

Our Corrolary 2 complements the Haug-Rousselet theorem on the existence of directional derivatives of eigenvalues at bifurcation points ([4]).

Chapter 6

References for Chapter 6

[1] Z. Nashed, The role of differentials, in "Nonlinear Functional Analysis and Applications" (L.B. Rall, Ed.), Proceedings of an advanced seminar & Math. Res. Center, Madison, Wisconsin, Academic Press, New York, (1971).

[2] B.N. Pschenichnyi, Necessary Conditions for an Extremum, Marcel-Dekker, New York, New York, 1971.

[3] E.J. Haug, K.K. Choi and V. Komkov, Design Sensitivity Analysis of Structural Systems, Academic Press, N.Y., 1986.

[4] E.J. Haug and B. Rousselet, Design Sensitivity Analysis in Structural Mechanics, II: Eigenvalue Variations, Journal of Structural Mechanics, Vol. 8, pp. 161-186, 1980.

[5] J.P. Zolesio, Semi-Derivatives of Repeated Eigenvalues, Optimization of Distributed Parameter Structures, Edited by E.J. Haug and J. Cea, Sijthoff and Noordhoff, Alphen aan den Rijn, Holland, pp. 1475-1491, 1981.

Appendix A

VARIATIONAL PRINCIPLES OF CONTINUUM MECHANICS VOLUME 2

Comments regarding the subtle difference between the function and its range.

Many problems of modern engineering or theoretical physics, particularly of mechanics, lead to evaluation of integrals along certain specific curves, or certain surfaces in two, or three dimensions. Let us suppose that we represent our problem in Cartesian 3-dimensional space with coordinates x, y, z. We denote the instantaneous position of a particle at time t by $\{x,y,z;\ t\}$. We also wish to describe the curve Γ transversed by the particle.

$$\Gamma:\quad z = z(x,y) \quad (1)$$

$$\text{or} \begin{cases} x = x(t), \\ y = y(t) \\ z = z(t) \end{cases} \quad (1^a)$$

would uniquely describe such curve, provided that either (1) or (1^a) is possible.

By now it is very difficult to distinguish what we mean by the symbol z.

What is the meaning of $\frac{dz}{dt}$, $\frac{\partial z}{\partial x}$, or dz?

If z is regarded as a coordinate representation in the system $\{x,y,z\}$ then $\frac{\partial z}{\partial y} = 0$, $\frac{\partial z}{\partial x} = 0$ and dz is not defined. Since x,y,z, t are independent variables (coordinates) then $\frac{dz}{dt}$ is either zero or else is undefined.

If we restrict our discussion to only two dimensions, we still encounter ambiguity which plagues specifically some areas of science such as thermodynamics, forcing many authors (including some historically famous and respected) to use notation that simply looks incompetent. The reason is clear. For example, pressure could be regarded as a dependent variable, as a coordinate of a thermodynamic system, or as a function of temperature and volume, or perhaps as a function of temperature only along a certain path in the p(v,T) space, with entrop y neglected, and soon. If someone writes p = p(t), where t is time, pressure is regarded as a function, but at the same time it may be an independent variable. Who knows what it is? It is the value of a function, the function itself regarded as a vector in a suitable Banach space, or only a coordinate? This clumsy notation forces authors to put additional brackets, explaining by subscripts or some other notation what does p(t) mean.

Let us introduce some notation that clearly distinguishes the value of a function from its numerical value, or whatever is a specific value of that function in its range. "Function" by itself is a meaningless word. The description using ordered pairs definition should specify the domain, the codomain and possibly some properties of the range (is it 1:1 onto,...?). Moreover one should describe the class of functions to which the function belongs, i.e. smoothness, closure properties, measurables,is it normable, does it belong to an inner product space?, etc., following the style of 20th century functional analysis, or general topology.

Membership in a class or space of functions may be denoted by an asterisk. (I am using the notation of Abraham Ungar [1]).

Hence $y(x) \underline{\underline{\text{def}}} \{x,(y(x)\}$, $y(x) \in R$, $x \in R$,

will be regarded as a number obtained by inserting x into the ordered pair $\{x,y(x)\}$ but $y^*(x)$ is a function whose properties must be specified, say $y^*(x) \in H$ where H is a Hilbert space, or $y^*(x) \in C^1 [0, 2\pi]$, etc... For example, $y = a^2 \in R$ is a number obtained by squaring a specific number $a \in \mathbb{R}$, it is not a function, it is a point in the range of $y^*(a)$, i.e. $\bar{y}^*(a) = a^2$, where $y^*(a): \mathbb{R} \to \mathbb{R}+$. We took the liberty of assigning a letter symbol y to a^2. We are free to do that.

$$\frac{dy}{da} = 0 \text{ since y and a are constants. But}$$

$$\frac{dy^*}{da} = 2a, \text{ if } y^*(a) = a^2,\ y^*: \mathbb{R} \to \mathbb{R} \ \in C^2(R).$$

A function $y^*(x)$ is an abstract quantity denoting all possible ordered pairs $\{x, y(x)\}$, $x \in \text{dom } y^*$. Without saying it, we assumed the simplest possible interpretation identifying the domain and range with some subset of the real numbers.

If we consider only the case when the domain and range is a subset of real numbers, we identify y to be a real number, y^* to be a function from $\mathbb{R} \to \mathbb{R}$, y^{**} to be a function from a set of functions ($\mathbb{R} \to \mathbb{R}$) into $\mathbb{R}$, and so on.

Suppose our function $y^*: \mathbb{R} \to \mathbb{R}$ is an element of a Hilbert space $\mathbb{H}$. Then y^{**} is a mapping from $\mathbb{H}$ into $\mathbb{R}$, y^{***} is a mapping from y^{**} into $\mathbb{R}$, and so on.

Let $\{x, y, \lambda\}$ denote points in $\mathbb{R}^3$. Then in our notation x,y,λ are numbers(independent of each other), $y^*(x,\lambda)$ is a function of two variables x, λ, $\lambda^*=\lambda^*(x,y)$ has the same meaning. Suppose that we define a plane in 3-space,with coordinates x,y,z,by the relation sloppily written as:$ax + by + cz = 0$. If $a \neq 0$ we could write $x^*(y,z) = 0^*(x,y,z)$, or $ax^*(y,z) = 0^*(x,y,z)$, where $0^*(x,y,z)$ is unique map from $\mathbb{R}^3 \to \mathbb{R}$ associating with every triple $\{x,y,z\}$ the number zero, while the value of $x^*(y,z)$, i.e. $\bar{x}^*$ is $(-by- cz)$. We shall denote the taking of the value of a function by a bar. Hence, $a \cdot \bar{x}^*(y,z) = - by - cz$ (It is legal to multiply a function by a constant). Similarly, we could solve the equation of the plane by regarding y as a function of x,z, defined by:$by^*(x,z) = 0^*(x,y,z)$, with $\overline{by^*}(x,z) = - ax - cz$.

In this notation x, z, a,c,b, are numbers

(in the domain Ω) while y^* is a function of the coordinates x and z. The letters $x,y,z...$ used for names of functions signify that the value of the function x^*, i.e. $\overline{x^*}$ is identified with the number denoted by x, otherwise we would use some "neutral" letters like $f(y,z)$ rather then $x^*(y,z)$.

This permits a generous use of the asterisk notation.

The equation $0 = ax + by + cz$ in our notation signifies equality of two numbers. The equation

$$ax^*(y,z) + by + cz = 0$$

makes no sense, because we are adding a function to two numbers. (And obtain a number?) Function can be added only to another function in the same space.

On the other hand

$\overline{ax^*(y,z)} + by + cz = 0$ makes sense. The equation

$$ax^*(y,z) + by^*(x,z) + c\,z^*(x,y) = 0^*(x,y,z)$$

is a true functional equation, which is satisfied by many functions not necessarily describing a plane in $\mathbb{R}^3$.

Let us consider sets of pairs $\{x,y\} \in \mathbb{R}^2$. For a functional relationship:

$y^*(x)$, $y^* \in \mathbb{H}$, $x \in \mathbb{R}$, if it is true that

$\frac{dy^*}{dx} \neq 0$ in the domain $a \leq x \leq b$, then we can define

$x^*(y)$, and the function $x^{**}(y^*(x))$. We require consistent use of the letter x, so that $x^{**}(y^*(x)) = x^*(\overline{y^*(x)})$, and $x^*(\overline{y^*(x)}) = x$.

Hence,

$$\overline{x^{**}(y^*(x))} = \overline{x^*\ (\overline{y^*(x)})} = x.$$

What does $\frac{dy^{**}}{dx^*}$ mean?

Can we clarify the meaning of differentiation of functions with respect to other functions?

As usual, we can define

$\dot{\lambda}^{**} = \frac{d\lambda^{**}}{d\mu^*}$ to be the linear operator defined by the relation

$$\lambda^{**}(\mu^* + t\Delta\mu^*) - \lambda^{**}(\mu^*)$$

$$= t(\dot{\lambda}^{**}\Delta\mu^*) + 0(t^2),$$ for sufficiently small $t\Delta\mu^*$, or simply for a sufficiently small t.

However, $\overline{\dot{\lambda}^{**}(\mu^*(x))} = \overline{\frac{d\lambda^*(\mu)}{d\mu}}$.

Hence, $\overline{\dot{\lambda}^{**}(\mu^*)} \quad \overline{\dot{\mu}^{**}(\lambda^*)}$ is defined and is

$$\dot{\lambda}^*(\mu) \cdot \dot{\mu}^*(\lambda) = 1.$$

Therefore $\dot{\lambda}^{**}$ is an object in an abstract space (it is certainly not a number!), but its evaluation obeys very simple rules. (See [3]) See [2] for a rigorous introduction.

References

[1] A. Ungar, A new operational calculus method, Int. J. of Engl Sc. Vol. 18, #1, (1980) p. 43-59.

[2] Z. Nashed, Differentiability and related properties of non-linear operators...in Non-linear functional analysis, L.B. Rall editor, Academic Press, New York, 1972, p. 103-309.

[3] Z. Nashed, The role of differentials, in "Nonlinear Functional Analysis and Applications"(L.B. Rall, Ed.), Proceedings of an advanced seminar & Math. Res. Center, Madison, Wisconsin, Academic Press, New York, (1971).

Appendix B

Lie derivatives in the setting of continuum mechanics

Let x^i be a coordinate system on a differentiable manifold V_n. The partial derivatives $\partial_i = \frac{\partial}{\partial x^i}$ form **basis** of a vector space $*T_x$ (See for example A. Lichnerowicz [1]. Since $df = \sum_i \frac{\partial f}{\partial x^i} dx^i$,

the vectors $\{dx^i\}$ forms basis of the space T_x dual to $*T_x$ with basis $\partial/\partial x^i$. The accepted notation that uses the summation convention is

$$df = \partial_i f \quad dx^i$$

We refer to elements of $*T_x$ as covariant vectors and to T_x as contravariant vectors.

As we change coordinate systems $x^{i'} = f^{i'}(x^i)$

we obtain $dx^{i'} = \frac{\partial f^{i'}}{\partial x^j} dx^j = A^{i'}_j \ dx^j$.

Thus, we can introduce arbitrary elements ω^α of $*T_x$ such that

$\omega^\alpha = a^\alpha_i \ dx^i$.

Dual basis Π_i, e, are defined by the relations

$\langle dx^i, \ \Pi_j \rangle = \delta^i_j$, $\langle \omega^\alpha, \ e_\beta \rangle = \delta^\alpha_\beta$, where δ^i_j is the

Kronecker's delta symbol

$$\delta^i_j \begin{cases} = 1 \text{ if } i = j \\ = 0 \text{ if } i \neq j. \end{cases}$$

The linear connection A^i_j satisfies

$$\partial_i A^\mu_\nu - \partial_\nu A^\mu_i = 0 \;, \; \mu = 1,2,\ldots n.$$

All transformations corresponding to changes of basis form a group:

$$e_{i'} = A^j_{i'} e_j \quad , \quad \omega^{i'} = A^{i'}_j \omega^j .$$

Convariant derivative ∇ has the following properties:

a) It is a linear operator
b) Application of ∇ to a scalar function f transforms f into a covariant vector ∇f which may be also denoted by *df.
c) It raises the rank of any tensor by one contravariant index(in the usual engineering interpretation of "tensors")
d) It obeys the Leibniz algebraic law of derivations.

$$\nabla(c_\alpha \omega^\alpha) = (\nabla c_\alpha) \cdot \omega^\alpha + c_\alpha \; (\nabla \omega^\alpha)$$

with

$$\nabla \omega^\alpha = \Gamma^\alpha_{\beta j} \; \omega^j \otimes \omega^\beta .$$

$\Gamma^\alpha_{\beta j}$ is the linear connection known as the

Christoffel symbol. It satisfies the conjugation law

$$\Gamma^\alpha_{\beta j} = \overset{*}{-}\Gamma^\alpha_{j\beta} \; .$$

We assert that $\nabla \omega^\alpha$ is a tensor of rank two. The concept of a deformation or of a moving coordinate system is closely related to the concept of an infinitesimal generator of a transformation and the corresponding Lie derivative.

A transformation assigning to a point

$\{x^i\} = p$, the point $p' = \{x^{i'}\} = \{x^i + \xi^i \cdot \delta t\}$

can be regarded as an infinitesimal transformation in which $\underset{\sim}{\xi}$ is the infinitesimal generator of a semigroup or as a process of dragging of a coordinate system through an infinitesimal vector

$dp = \langle dx^i, \Pi_i \rangle = \xi^i \, \mathbf{e}_i \, \delta t.$

The "natural" coordinate system $\{\mathbf{e}_i\}$ is defined in [2] by V.E. Stepanov and Iu. A. Iappa

<u>A definition of the Lie derivative.</u>

The Lie derivative of a vector $\underset{\sim}{u}$ in the direction of a vector η can be defined as the commutator

$L_\eta \underset{\sim}{u} = \langle \eta, \nabla \underset{\sim}{u} \rangle - \langle \underset{\sim}{u}, \nabla \eta \rangle$

The Lie derivative of a scalar function f(x) is defined by the rule

$L_\eta f = \eta^i \cdot \partial_i f.$

Hence, it is the directional derivative, that is a projection of gradient of f in the direction of η.

For a vector $\underset{\sim}{u} = c^i \, \mathbf{e}_i$, where $\mathbf{e}_i$ are basis, the Lie derivative is defined by Leibniz' rule.

$L_\eta \underset{\sim}{u} = L_\eta (c^i \mathbf{e}_i) = (L_\eta c^i) \, \mathbf{e}_i + c^i (L_\eta \mathbf{e}_i)$

$= (\eta^i \partial_i \, c^j) \, \mathbf{e}_j + c^i (L_\eta \, \mathbf{e}_i)$

We recall that

$L_\eta \mathbf{e}_i = \eta^j \, \Gamma^k_{ij} \, \mathbf{e}_k - \nabla_i \, \eta^j \, \mathbf{e}_j.$

Let $\underset{\sim}{u}$ be a vector expressed in a moving coordinate system. Let $\mathbf{e}_i$ be local basis. Then

$$\mathbf{e}_i(\underset{\sim}{x} + d\underset{\sim}{x}) = \mathbf{e}_i(\underset{\sim}{x}) + \Gamma^j_{ik} \, \eta^k \, \mathbf{e}_j \cdot \delta t$$

Lie derivatives are defined by

$$L_\eta u^i\, \delta t = u^i(x + dx) - \tilde{u}^i(x + dx)$$

where $\tilde{u}^i$ is the Lagrangian representation of a vector dragged along a moving coordinate system

$$\tilde{u}^i(\underset{\sim}{x} + d\underset{\sim}{x}) = u^i(\underset{\sim}{x}) + \Omega^i_j\, u^j(\underset{\sim}{x})\cdot\delta t\ \cdot$$

with $\Omega^i_j = \nabla_j \eta^i - \Gamma^i_{jk} \eta^k$. where Γ^i_{jk} are the Christoffel symbols.

Bibliography

[1] A. Lichnerowicz, Geometrie de Groupes des Transformations, Dunod, Paris, 1958.

[2] V.E. Stepanov and Iu. A. Iappa, Lie derivative for geometric objects, Vestnik Leningrad. Univ. #22(1978), p. 42-46.

Index